SCOTLAND'S CULTURAL HERITAGE

VOLUME IV

THE ROYAL SOCIETY OF EDINBURGH:
100 MEDICAL FELLOWS ELECTED 1841-1882

An illustrated catalogue containing a brief biographical sketch of each Fellow and references to publications, by and about him as well as the locations of portraits, prints, busts, manuscripts etc.

Compiled by
Flora Bennet, Sheila Devlin-Thorp, Maureen Doran, Steven Garrad,
Tracy MacAlpine, Rosemary Moodie and Sigridur Oladottir.

Edited by
Anne Boyle, Timothy Hayward and Laurence Hunter

"Scotland's Cultural Heritage" is a Manpower Services Commission (CEP) Project sponsored by Professor Eric G. Forbes and based at the History of Medicine and Science Unit, High School Yards, Drummond Street, Edinburgh EH1 1LZ. It is administered by the University of Edinburgh.

Published by
Scotland's Cultural Heritage,
History of Medicine and Science Unit,
The University of Edinburgh,
High School Yards,
Edinburgh
EH1 1LZ

ISBN   0 907692 11 7

INTRODUCTION

This is the fourth volume in the "Scotland's Cultural Heritage" series of catalogues of Fellows of the Royal Society of Edinburgh, elected between 1783 and 1882. It covers 100 Fellows which is the majority of medical Fellows elected between 1841 and 1882. Many of these are well-known figures but inevitably it also includes some less prominent individuals who have required considerable amounts of research to establish what is reproduced here. Volume 3 covered the earlier period of medical Fellows elected between 1783 and 1844.

As with other volumes, the format used has been chosen so as to allow as much information as possible about each Fellow to be detailed on a single A4 size page with a photograph included whenever one is available. It was therefore necessary to be selective in those cases where a considerable amount of information is at hand. However such deliberate omissions have been indicated by an asterisk (*) after the appropriate sub-heading, priority always being given to the most relevant details. Thus, for example, if there are numerous entries under <u>"Portraits, etc."</u>, group pictures would be omitted.

For manuscript locations we have checked immediately prior to publication the National Register of Archives (NRA). Any enquiries relating to private collections in England should be addressed in the first instance to Mrs S. Wilmington, NRA, Quality Court, Chancery Lane, London; while those pertaining to Scotland should be addressed to Miss D.M. Hunter, West Register House, Scottish Record Office, Edinburgh EH2.

In the section which is entitled <u>"Portraits, etc.,"</u> a double asterisk (**) denotes which portrait has been used for the accompanying illustration. Every effort was made to establish the locations of portraits of the Fellows concerned but we do not pretend that the list is complete. Scotland's Cultural Heritage will be pleased to receive information on the nature and location of any other relevant works.

In this volume, the biographical material was collected by Maureen Doran, Sheila Devlin-Thorp and Sigridur Oladottir. Anne Boyle and Timothy Hayward undertook the checking and editing of the material as it was produced for publication. Steven Garrad prepared the illustrations and with the assistance of Tracy McAlpine and Flora Bennet, located the portraits, engravings etc. Rosemary Moodie and Timothy Hayward were responsible for the typing and processing of the text on a Superbrain QD computer which has been acquired together with a printer by "Scotland's Cultural Heritage" with the aid of generous grants from the University of Edinburgh and the Manpower Services Commission.

We are most grateful to the Manpower Services Commission for their contin--uing support to the project.

Laurence J.K. Hunter
11 March 1983

<u>ACKNOWLEDGEMENTS</u>

Scotland's Cultural Heritage is grateful for the valuable assistance given in the preparation of this catalogue by Miss E.S. Cumming, Miss J. Ferguson, Miss S.D. Fletcher, Mrs R. Gibson, Mr M. Nicolson and Mr J. Rock.

Also we wish to thank the staffs of the:

British Library, London

Historical Manuscripts Commission, London

National Portrait Gallery, London

National Register of Archives, London

Public Records Office, London

Wellcome Institute, London

and the following Edinburgh libraries, museums, galleries and archives:

Edinburgh Public Library

Edinburgh University Library

General Register House

Edinburgh Royal Infirmary Medical Archives Centre

National Gallery of Scotland

National Library of Scotland

Royal Botanical Gardens, Edinburgh

Royal College of Physicians of Edinburgh

Royal College of Surgeons of Edinburgh

Royal (Dick) School of Veterinary Studies

Royal Scottish Academy

Scottish National Portrait Gallery

West Register House

Although no official endorsement for this publication has been sought from any of these institutions, we are most grateful to them for the information and advise they have generously given in response to our enquiries.

The team accepts full responsibility for any factual errors, omissions or misrepresentations that inadvertently may have been incurred in the course of producing this publication.

Scotland's Cultural Heritage is most grateful to the Department of Geography, the Research Centre for Social Sciences and the Central Administration of the University of Edinburgh for their assistance to this project and its work.

<u>Illustrations:</u>

We are grateful to the following for supplying and allowing us to reproduce their portraits and photographs:

The National Portrait Gallery for portraits of Allman, Balfour, Sir J. Browne, Fayrer, Mitchell and Symonds.

The Trustees of the National Library of Scotland for portraits of J. Anderson, Cormack, Craig, J. Gamgee, Moreland, W.B. Thomson and Wright.

Edinburgh University Library for portrait of T. Anderson.

The Royal College of Physicians of Edinburgh for portraits of J. Begbie, Moir, Peddie, Seller and Tuke.

The Medical Archives Centre of the Royal Infirmary of Edinburgh for portraits of John Duncan, Haldane, Lister, Macleod and Wood.

The Royal Infirmary of Glasgow for portrait of J. Fleming.

The University of Edinburgh for portraits of Maclagan and Young.

The Royal (Dick) School of Veterinary Studies for portrait of Williams.

The Royal College of Physicians for portrait of Wylie.

The William Art Gallery and Museum, Birkenhead for portrait of Higgins.

All the other illustrations are reproduced by kind permission of the Wellcome Institute from their invaluable publication 'History of Scottish Medicine' by J.D. Comrie (London, 1932).

We are grateful to the Council of the Royal Society of Edinburgh for granting permission to reproduce their crest as the frontispiece illustration.

ABBREVIATIONS

Manuscript and Botanical Collections

All of the symbols in this section relating to manuscript locations are (with some modifications) based on the system used in the British Union Catalogue of Periodicals, (Eds. James D. Stewart; Muriel E. Hammond; E. Saenger - London 1955), issued annually since 1955.

Botanical collections are identified by (IH) following the symbols, which are based on the system used in the 'Index Herbarium; Part 1. The Herbaria of the World, 6th edition'. Compiled by P.K. Holmgren and W. Keuken. (October 1974; Utrecht, Netherlands)

BM(IH) British Museum (Natural History), Cromwell Road, LONDON SW7 5BD

BU University of Birmingham Library, Edgbaston, BIRMINGHAM, B15 2TT

CSR Scott Polar Research Institute, Lensfield Road, University of Cambridge, CAMBRIDGE

E National Library of Scotland, George IV Bridge, EDINBURGH EH1

EF Royal Scottish Museum, Chambers Street, EDINBURGH EH1 1JF

E(IH) Royal Botanical Gardens, Inverleith Row, EDINBURGH EH3 5LR

ERM Royal Medical Society of Edinburgh, Students Centre, Bristo Street, EDINBURGH

EPH Royal College of Physicians of Edinburgh, 9 Queen Street, EDINBURGH EH2 1JQ

ES Royal College of Surgeons of Edinburgh, 18 Nicolson Street, EDINBURGH EH8

EU University of Edinburgh Library, George Square, EDINBURGH EH8 9LQ

GC Strathclyde Regional Archives, 30 John Street, GLASGOW G2 4PF

GF Royal College of Physicians and Surgeons, 242 St. Vincent Street, GLASGOW G2

GU University of Glasgow Library, Hillhead Street, GLASGOW G12 8QE

KB Royal Botanic Gardens, KEW, Richmond, Surrey TW9 3AE

K(IH) The Herbarium and Library, Royal Botanic Gardens, KEW, Richmond, Surrey TW9 3AE

KM Commonwealth Mycological Institute, Ferry Lane, KEW, Surrey

L The British Library, Great Russell Street, LONDON WC1E 7DG

LA Royal Army Medical College, Millbank, LONDON SW1

LCP Royal College of Physicians of London, 11 St. Andrews Place, LONDON NW1 4LE

LGS Royal Geographical Society, 1 Kensington Gore, LONDON SW7

LIC Imperial College of Science & Technology Archives, Sherfield Building, Imperial College, LONDON SW7 2AZ

LINN(IH) Linnean Society of London, Burlington House, Piccadilly, LONDON W1V 0LQ

LLS Linnean Society of London, Burlington House, Piccadilly, LONDON W1V 0LQ

LR The Royal Society of London, 6 Carlton House Terrace, LONDON SW1

LS Royal College of Surgeons of England, 35-43 Lincoln's Inn Fields, LONDON WC2A 3PN

LUC University College London, DMS Watson Library, Gower Street, LONDON WC1E 6BT

LWM Wellcome Institute for the History of Medicine, 183 Euston Road, LONDON NW1 2BP

MU The John Rylands University Library of Manchester, Oxford Road, MANCHESTER M13 9PP

O Bodleian Library, Broad Street, OXFORD OX1 3BG

SaU University of St. Andrews Library, ST. ANDREWS, Fife

Other abbreviations:

These abbreviations and symbols have been used in parts of the text other than the manuscript locations section.

* Only a part of the information available has been noted.

** Denotes which of the portraits or illustrations listed has been reproduced.

n.p. No place of publication

n.d. Date unknown

BMC British Museum Catalogue

DNB Dictionary of National Biography

DSB Dictionary of Scientific Biography

MSS.Cat. EUL Catalogue of Manuscripts in the University of Edinburgh Library

NUC National Union Catalogue

NPG National Portrait Gallery

RCP Royal College of Physicians, London

RCPE Royal College of Physicians of Edinburgh

SNPG Scottish National Portrait Gallery

INFORMATION SOURCES

During the preparation of this catalogue over 300 biographical and reference works have been consulted. Among the principle sources used were:

Biographisches Lexikon der Hervorragenden Aerzte   (Berlin, 1929)
Boase, Frederic, Modern English Biography, 3 vols., 2nd edition (London, 1965)
British Museum Catalogue of Printed Books (London, 1964)
Britten, J. and Boulger, G.S., A Biographical index of deceased British and Irish Botanists, 2nd edition (London, 1931)
Chambers, R., Dictionary of Eminent Scotsmen (London, 1875)
Comrie, J.D., History of Scottish Medicine, 2 vols. (London, 1932)
Crawford, D.G., History of the Indian Medical Service (London, 1914)
Crawford, D.G., Roll of the Indian Medical Service, 1615–1930 (London, 1930)
Desmond, R., Dictionary of British and Irish Botanists and Horticulturalists (London, 1977)
Eddington, A., Edinburgh and Lothians at the opening of the Twentieth Century (Brighton and Edinburgh, 1904)
Gillespie, C.C., The Dictionary of Scientific Biography, 14 vols. (New York, 1970-76)
Gray, J., History of the Royal Medical Society 1737-1937 (Edinburgh, 1952)
Lee, Sir S. and Stephen, Sir L., The Dictionary of National Biography, 21 vols, various editions.
Leyland, J., Contemporary Medical Men (Leicester, 1888)
Logan-Turner, A., History of the University of Edinburgh 1833-1933 (London, 1933)
Munk, W., The Roll of the Royal College of Physicians of London, 3 vols. (London, 1878)
The National Union Catalogue. Pre-1956 Imprints (London and New York, 1968)
Plarr, V.G., (revised by Sir D'Arcy Power), Lives of the Fellows of the Royal College of Surgeons of England (London, 1930)
Royal College of Physicians of London. Library Catalogue (London, 1912)
Who's Who, various editions.
Who Was Who, various editions.

Periodicals, etc.

(The abbreviations used follow, with modifications, the system used by Brown, P. and Stratton, C.B. eds., World List of Scientific Periodicals)

Br. med. J.   British Medical Journal
Br. vet. J.   British Veterinary Journal
Bull. Hist. Med.   Bulletin of the History of Medicine
Caledon. med. J.   Caledonian Medical Journal
Edinb. Med. J.   Edinburgh Medical Journal
Edinb. Med. Surg. J.   Edinburgh Medical and Surgical Journal
Edinb. New Phil. J.   Edinburgh New Philosophical Journal
Illus. Lond. News   Illustrated London News
J. Roy. Nav. Med. Serv.   Journal of the Royal Navy Medical Service
J. Soc. Biblphy. nat. Hist.   Journal of the Society for the Bibliography of Natural History
Phil. Trans. R. Soc.   Philosophical Transactions of the Royal Socierty of London
Proc. Linn. Soc.   Proceedings of the Linnean Society of London
Proc. R. phil. Soc. Glasg.   Proceedings of the Royal Philosophical Society of Glasgow
Proc. R. Instn. Gt. Br.   Proceedings of the Royal Institution of Great Britain
Proc. R. Soc. Edinb.   Proceedings of the Royal Society of Edinburgh
Trans. Bot. Soc. Edinb.   Transactions of the Botanical Society of Edinburgh
Trans. med.-chir. Soc. Edinb.   Transactions of the Medico-Chirurgical Society of Edinburgh
Trans. obstet. Soc. Lond.   Transactions of the Obstetrical Society of London
Trans. R. Scott. Soc. Arts.   Transactions of the Royal Scottish Society of Arts
Trans. R. Soc. Edinb.   Transactions of the Royal Society of Edinburgh
Vet. J.   The Veterinary Journal

<u>VOLUME FOUR</u> - <u>CONTENTS</u>

CRICHTON-BROWNE, Sir James   (1840-1938)
Physician and Psychologist

CUNYNGHAME, Robert James Blair   (1841-1903)
Physician, Superintendent of Statistics

DUNCAN, James   (1810-1866)
Senior Surgeon to the Royal Infirmary

DUNCAN, James Matthews   (1826-1890)
Physician-Accoucheur at St. Bartholomew's Hospital, London

DUNCAN, John   (1839-1899)
Surgeon

DUNSMURE, James   (1814-1886)
Surgeon

DYCE, Robert   (1798-1869)
Professor of Midwifery at the University of Aberdeen

EDWARDS, Alexander Mackenzie   (-1868)
Surgeon

FAYRER, Sir Joseph   (1824-1907)
Physician Extra-ordinary to the King

FLEMING, Andrew   (1822-1901)
Surgeon and Naturalist

FLEMING, John Gibson   (1809-1879)
Physician and Surgeon

FRASER, Sir Thomas Richard   (1841-1920)
Pharmacologist

GAMGEE, Arthur   (1841-1909)
Physician and Physiologist

GAMGEE, Joseph Sampson   (1828-1886)
Surgeon, Queen's Hospital, Birmingham

GLOVER, Robert Mortimer   (1816-1859)
Physician

GRAHAM, Andrew   (1817-1889)
Fleet Surgeon, Royal Navy

HALDANE, Daniel Rutherford   (1824-1887)
Physician

HALLEN, James Herbert Brockencote   (1829-1901)
Inspecting Veterinary Surgeon to H.M. Indian Army

HERAPATH, William Bird   (1820-1868)
Surgeon and Toxicologist

HIGGINS, Charles Hayes   (1811-1898)
Surgeon

HUNTER, Alexander   (1816-1890)
Surgeon-Major, Madras Army

KEILLER, Alexander   (1811-1892)
Gynaecologist

LAYCOCK, Thomas   (1812-1876)
Mental Physiologist, Professor of the Practice of Physic at the University of Edinburgh

LINDSAY, William Lauder   (1829-1880)
Alienist and Botanist

LISTER, Lord Joseph   (1827-1912)
Professor of Clinical Surgery, King's College, London

LOGIE, Cosmo Gordon   (1820-1886)
Surgeon Major of the Royal Horse Guards

LOWE, William Henry   (1814-1900)
Physician

McBAIN, James   (1807-1879)
Surgeon and Naturalist in the Royal Navy

MACCALL, Thomas Smith   (c.1804-1895)
Physician

MACDONALD, Angus   (1836-1886)
Obstetrician

McKENDRICK, John Gray   (1841-1926)
Professor of Physiology, Emeritus Professor at the University of Glasgow

MacLAGAN, Sir Andrew Douglas   (1812-1900)
Professor of Medical Jurisprudence and Public Health at the University of Edinburgh (1862-1897)

MacLAGAN, Robert Craig   (1839-1919)
Physician

MacLEOD, Sir George Husband Baird   (1828-1892)
Regius Professor of Surgery at the University of Glasgow

MALCOLM, Robert Bowes   (1808-1894)
Physician

MITCHELL, Sir Arthur   (1826-1909)
Commissioner in Lunacy; Antiquary

MOIR, John   (1808-1899)
Professor of Midwifery at the University of Edinburgh

MOREHEAD, Charles   (1807-1882)
Physician in the Bombay Medical Service

MURRAY, John Ivor   (1824-1903)
Physician

NAYSMITH, Robert   (1792-1870)
Surgeon and Dentist

NEWBIGGING, Patrick Small Keir   (1813-1864)
Physician

PAGAN, Samuel Alexander   (-   )
Physician

PEDDIE, Alexander   (1810-1907)
Physician

PETTIGREW, James Bell   (1834-1908)
Chandos Professor of Medicine and Anatomy at the University of St. Andrews

PIRRIE, William   (1807-1882)
Regius Professor of Surgery, Marischal College, Aberdeen

RICHARDSON, Sir John   (1787-1865)
Surgeon, Arctic Explorer and Naturalist

ROBERTSON, Douglas Argyll   (1837-1909)
Ophthalmic Surgeon

ROBERTSON, William   (1818-1882)
Physician

RUTHERFORD, William   (1839-1899)
Professor of the Institutes of Medicine at the University of Edinburgh

SANDERS, William Rutherford   (1828-1881)
Physician; Professor of Pathology at the University of Edinburgh

SANDERSON, James   (1812-1891)
Surgeon

SCOTT, John   (1797-1859)
Physician

SELLER, William   (1798-1869)
Physician and Botanist

SHEARMAN, Edward James   (1798-1878)
Physician

SIBBALD, Sir John   (1833-1905)
Commissioner in Lunacy for Scotland

SIMPSON, Alexander Russell   (1835-1916)
Professor of Midwifery at the University of Edinburgh

SMITH, John   (1798-1879)
Physician

SPENCE, James   (1812-1882)
Professor of Surgery at the University of Edinburgh

SPITTAL, Robert   (1804-1852)
Physician

STARK, James   (1811-1890)
Physician; Superintendent of Medical Statistics for Scotland

STEWART, John Lindsay   (1831-1873)
Surgeon and Botanist

STEWART, Sir Thomas Grainger   (1837-1900)
Professor of the Practice of Physic at the University of Edinburgh

SYMONDS, John Addington   (1807-1871)
Physician to the General Hospital and Lecturer on Forensic Medicine, Bristol

TAYLOR, Sir Alexander   (1790-1879)
Physician

THOMSON, Allen   (1809-1884)
Biologist; Emeritus Professor of Anatomy at the University of Glasgow

THOMSON, William Burns   (1821-1893)
Medical Missionary

TUKE, Sir John Batty   (1835-1913)
Alienist and Medical Writer

TURNER, Sir William   (1832-1916)
Principal and Vice-Chancellor of the University of Edinburgh

WATSON, James   (1792-1877)
Physician

WATSON, Morrison   (1845-1885)
Professor of Anatomy at Owen's College, Manchester

WATSON, Sir Patrick Heron   (1832-1907)
Consulting Surgeon to Chalmers Hospital, Edinburgh

WILLIAMS, William   (1832-1900)
Principal and Professor of Veterinary Medicine and Surgery, New Veterinary College, Edinburgh

WILLIAMSON, Thomas   (1815-1885)
Surgeon at Leith Hospital

WILSON, James George   (1830-1881)
Professor of Midwifery at Anderson College, Glasgow

WISE, Thomas Alexander   (1802-1889)
Principal of Dakka College, India

WOOD, Alexander   (1817-1884)
Physician

WOOD, Andrew   (1810-1881)
Physician

WRIGHT, Thomas   (1809-1884)
Physician; palaeontologist; geologist

WYLIE, John   (1790-1852)
Physician-General in the Madras Army

YOUNG, John   (1835-1902)
Professor of Natural History at the University of Glasgow

VOLUME IV

THE ROYAL SOCIETY OF EDINBURGH

100 MEDICAL FELLOWS ELECTED 1841-1882

George James ALLMAN   (1812-1898)

Regius Professor of Natural History at the University of Edinburgh

Elected F.R.S.E. 21 January 1856

Portraits, etc.

Chalk drawing: (F.W.Burton), National Gallery of
    Ireland, Dublin
Portrait: (E.M.Buck, 1898), Linnean Society,
    London
Bust: (John Hutchinson, RSA, 1862), University of
    Edinburgh
Lithograph: (T.H.Maguire, 1851), NPG **

Born in Cork, Ireland, in 1812, the son of James Allman, George received his early education at the Belfast Academical Institution. He then went to Trinity College, Dublin with a view to entering the legal profession but after graduating B.A. in 1839, decided on a scientific career and so took his medical degree four years later. Soon after he was appointed Professor of Botany at Dublin University in succession to William Allman. In 1847 he graduated M.D. from Oxford and eight years later left Dublin to become Regius Professor of Natural History at Edinburgh University. He was also appointed keeper of the city's natural history museum. He resigned from the professorship in 1870 on account of ill-health and retired to Parkstone in Dorset where he devoted his time to zoological research and his favourite hobby of horticulture. He died at Parkstone on 24 November 1898.

Honours and Distinctions *

1854   Fellow of the Royal Society of London
1872   Fellow of the Linnean Society
1873   Gold Medal of the Royal Society of London
1874   President of the Linnean Society
1877   Brisbane Medal of the Royal Society of London
1878   Cunningham Medal of the Royal Irish Academy
1879   President of the British Association for the Advancement of Science

Publications *

Introductory lecture delivered to the students of the natural history class in the University of Edinburgh (Edinburgh, 1855)
Monograph of the freshwater polyzoa (London, 1856)
The method and aim of natural history studies (Edinburgh, 1868)
A monograph of the gymnoblastic on tubularian hydroids (London, 1871)

Biographical Studies *

DNB, (London, 1968), supplement
Desmond, R., Dictionary of British and Irish Botanists and Horticulturalists (London 1977), p.9

Manuscript Location

E; EU; LIC; LR; SaU

Collections

LINN (IH)

John ANDERSON (1833-1900)

Professor of Comparative Anatomy at the Medical
College, Calcutta

Elected F.R.S.E. 12 April 1871

Portraits, etc.

None known

Born in Edinburgh on 4 October 1833, the son of
Thomas Anderson, secretary to the National Bank
of Scotland, John received his early education
at the Edinburgh Institution. He then studied
medicine at Edinburgh University, graduating
M.D. in 1861 and was awarded the gold medal for
his outstanding performance. Two years later he
was appointed Professor of Natural Science at
the Free Church College in Edinburgh but left
this post after only one year to take up the
Professorship of Comparative Anatomy at the
Medical College of Calcutta. He also held the post of Superintendent of the Indian Museum in
Calcutta and accompanied two expeditions to Yunnan in Western China as scientific officer. He
later published several reports on his findings. In 1886 he retired from the Indian Medical
Service and returned to London. He died at Buxton, Derbyshire on 15 August 1900. He was married
to Grace Scott, daughter of Patrick Hunter Thoms of Aberlemno in Forfarshire.

Honours and Distinctions

1879   Fellow of the Royal Society of London
1885   Fellow of the Royal Geographical Society
n.d.   Fellow of the Linnean Society
n.d.   Fellow of the Society of Antiquaries

Publications

A report on the expedition to Western Yunnan via Bhamo (Calcutta, 1871)
Mandalay to Momien: a narrative of two expeditions to Western China of 1868 and 1875
(London, 1876)
"Anatomical and zoological researches" (London and Calcutta, 1878)
Catalogue of mammalia (Calcutta, 1881)
Handbook to the archaeological collections of the Indian Museum, Calcutta, (Calcutta, 1883-4)
English intercourse with Siam in the 17th century, (London, 1890)
A contribution to the herpetology of Arabia (London, 1896)
Zoology of Egypt, 2 Vols (London, 1898-1907)

Biographical Studies

DNB, (London, 1968), supplement
Proc. R. Soc. Lond., 75, (London, 1905), pp.113-116
Nature, 62, (London, 1900), pp.529-531
Science, 12, (London, 1900), pp.379-380

Manuscript Locations

None known

John ANDERSON (1840-1910)

Surgeon

Elected F.R.S.E.   2 March 1874

<u>Portraits, etc.</u>

Photograph: see The Lancet, 1910 **

John Anderson was born on 18 January 1840, the son of Robert Anderson. He received his medical education at the Royal Infirmary, Manchester where he was a clinical clerk and dresser to one of the senior surgeons. In 1861 Anderson became a member of the Royal College of Surgeons of England and was appointed successively house surgeon and house physician at the Royal Infirmary, Manchester. He entered the Army Medical Service in 1864 serving many years in India in the Royal Artillery until the regimental system was abolished. Returning home Anderson served in Edinburgh and at Camberley in connection with the Royal Military College. While in Edinburgh he went through a course of practical anatomy at Edinburgh University. Anderson went back to India when selected by the Marquis of Ripon for his personal staff, returning to England at the end of Lord Ripon's career as Viceroy. He retired from the Army in 1885 with the rank of Brigade-Surgeon and began practice in London becoming physician to the Dreadnought Hospital for Seamen and to King Edward's Hospital for officers. Anderson was also appointed lecturer on tropical diseases at St. Mary's Hospital. When he decided to begin practice in London, Anderson graduated in medicine at the University of St. Andrews and became a member and subsequently a Fellow of the Royal College of Physicians of London. Anderson was married to Jessie Usher and had four children by her. He died in North Kensington, London on 10 October 1910.

<u>Honours</u> and <u>Distinctions</u>

1898   Fellow of the Royal College of Physicians, London

<u>Publications</u>

None known

<u>Biographical Studies</u>

Munk, W., Roll of the Royal College of Physicians, <u>4</u>, (London, 1878), p.384
<u>The Lancet,</u> (1910)

<u>Manuscript Locations</u>

None known

Thomas ANDERSON (1819-1874)

Regius Professor of Chemistry at the University
of Glasgow

Elected F.R.S.E. 3 February 1845

<u>Portraits, etc.</u>

Photograph: see Glasgow University Old and
New... (ed.), W Stewart, (Glasgow, 1891) **

Born at Leith, Edinburgh on 2 July 1819, the
son of a local physician, Thomas received his
early education at the High School and the
Edinburgh Academy. He then studied medicine at
Edinburgh University, graduating M.D. in 1841.
After further study abroad, which included
periods under Berzelius in Stockholm, and
Liebig in Giessen, he returned to Edinburgh and
in 1846 began to lecture in the extra-mural
school. In 1852 he succeeded Thomas Thomson as
Professor of Chemistry at Glasgow University. His most important work was in the field of
organic chemistry, particularly his discovery of picoline which he found was an isomer of
anilene, being the first discovered member of the pyridine series of bases. He later discovered
pyridine itself and its methyl derivation through research on the distillation of bone oil. He
concluded from his discovery of pyridine, lutidine and collidine that these formed a homologous
series and were derived from 'ammonia by the replacement of its three atoms of hydrogen by as
many different radicals'. He also made a detailed investigation of codeine and other opium
constituents, elucidating the composition of a number of alkaloids. He also found the true
constitution of anthracene. As chemist to the Highland and Agricultural Society of Scotland he
did research on soils, manures, and the composition of wheat. His findings were published over a
period of twenty-five years. For a time he also edited the 'Edinburgh New Philosophical
Journal'. In 1869 his health began to deteriorate rapidly and paralysis and deafness meant he
could no longer work. He died at Chiswick, London on 2 November 1874.

<u>Honours and Distinctions</u>

1872   Royal Medal of the Royal Society of London
1876   Fellow of the Royal College of Physicians of London

<u>Publications</u> *

On the constitution and properties of picoline, a new organic base from coal-tar
(Edinburgh, 1846)
Note on the constitution of the phosphates of the organic alkalites (London, 1849)
On the constitution of codeine and its products of decomposition (Edinburgh, 1853)
On the constitution of anthracene ... and some of its products of decomposition
(Edinburgh, 1861)

<u>Biographical Studies</u>

DSB, (New York, 1976)
Williams, T.I., (ed.), A Biographical Dictionary of Scientists (London, 1969), p.12

<u>Manuscript Locations</u>

EU; GU; SaU

James ANDREW (1811-1859)

Physician

Elected F.R.S.E. 6 January 1845

<u>Portraits, etc.</u>

None known

Born in the East India College at Addiscombe in Surrey, son of the Rev. James Anderson, governor of the College, James entered Caius College, Cambridge in 1828. He later studied medicine at Edinburgh University, but kept his terms at Cambridge and graduated M.D. there in 1839. In the same year he moved to Edinburgh and in due course was appointed one of the medical officers in the New Town Dispensary and the Royal Public Dispensary. In 1846 he was appointed ordinary physician to the Royal Infirmary, Edinburgh and remained there for the next ten years. Andrew was active in medical matters of the city, being at various times a manager of the Infirmary and the Morningside Lunatic Asylum, medical secretary to the Royal Dispensary, and at the time of his death, a member of the Council of the Royal College of Physicians. He died at the age of forty-eight at 15 Queen Street, Edinburgh on 1 December 1859, leaving a wife and six children.

<u>Honours and Distinctions</u>

1840   Fellow of the Royal College of Physicians of Edinburgh

<u>Publications</u>

Case of poisoning by corrosive sublimate   (Edinburgh, 1845)
Case of sudden death from rupture of muscular fibres of the heart   (n.p., 1845)
Case of injury of the pelvis during pregnancy   (n.p., 1850)
Case of poisoning by atropine   (Edinburgh, 1852)

<u>Biographical Studies</u>

Boase, F., Modern English Biography, <u>1</u>, (London 1965), p.67
<u>Edinb. Med. J.</u>, <u>5</u>, (Edinburgh, 1860), pp.674-675

<u>Manuscript Locations</u>

EU

Thomas ANNANDALE (1838-1907)

Professor of Clinical Surgery at the University
of Edinburgh

Elected F.R.S.E. 4 February 1867

Portraits, etc.

Bust: (W.G. Stevenson, R.S.A.), Edinburgh Royal
    Infirmary
Photograph: see Comrie, J.D., History of Scottish
    Medicine, (London, 1932) **

Born at Newcastle-on-Tyne on 2 February 1838,
the second son of Thomas Annandale, a surgeon,
Thomas received his early education at Bruce's
Academy, Newcastle, after which he was
apprenticed to his father. In 1856 he
matriculated at Edinburgh University and
graduated M.D. four years later with the
highest honours, receiving the gold medal for
his thesis 'On the injuries and diseases of the hip-joint'. In the same year he was appointed
house-surgeon to James Syme, and remained his private assistant until 1870. In 1863 he became a
junior demonstrator of anatomy in Edinburgh University under Professor John Goodsir, as well as
being appointed a lecturer on the principles of surgery in the extra-mural school of medicine.
Here he gave an annual course of lectures until 1871, when he began lecturing on clinical
surgery at the Edinbugh Royal Infirmary. Six years later he succeeded Lister as Regius Professor
of Clinical Surgery at the University, occupying this chair for thirty years. 'The Annandale
Gold Medal in Clinical Surgery', awarded annually by the Univeristy of Edinburgh, was founded in
his memory. From 1900 till death he was Surgeon-General to the Royal Company of Archers, having
joined the corps as an archer in 1870. Annandale was highly regarded as a clear lecturer and a
very skilled surgeon. He died suddenly in Edinburgh on 20 December 1907 and was buried in the
Dean Cemetery.

Honours and Distinctions

1863   Fellow of the Royal College of Surgeons of Edinburgh
1864   Fellow of the Royal College of Surgeons of England
1864   Awarded Jacksonian Prize by the Royal College of Surgeons of England
1902   Honorary D.L.C., Durham University

Publications *

Observations and cases in surgery   (Edinburgh, 1865)
Abstracts of Surgical Principles   (Edinburgh, 1868)
The treatment of congenital talipes virus in the infant   (Edinburgh, 1869)
On fatty hernia   (Edinburgh, 1870)
Observations on amputation at the hip-joint   (Edinburgh, 1870)
On the pathology and operative treatment of 'hip' disease   (Edinburgh, 1876)
Surgical appliances and minor operative surgery   (Edinburgh, 1866)

Biographical Studies

DNB., (London, 1912), Second supplement
Gray, J., History of the Royal Medical Society 1737-1937, (Edinburgh 1952), pp.236-8
Who's Who 1898, (London, 1898)
Comrie, J.D., History of Scottish Medicine, 2, (London 1932), pp.629, 670-3, 712

Manuscript Locations

E; EU

Thomas Graham BALFOUR (1813-1891)

Surgeon-General in the Army

Elected F.R.S.E. 17 January 1870

<u>Portraits, etc.</u>

Etching: (W.Strang), NPG **

Born in Edinburgh on 18 March 1813, Thomas received his early education at the Royal High School and the Edinburgh Academy. He then went to Edinburgh University to study medicine and graduated M.D. in 1834. Two years later he joined the army, his first task being to assist in clarifying medical statistics. Six years later he was appointed assistant surgeon to the Grenadier Guards. He served with them for eight years, when he became surgeon to the Duke of York's Asylum for Soldiers' Orphans at Chelsea. In 1857 Balfour was appointed as secretary to Sidney Herbert's commission, whose reports completely re-organized the Army Medical Department. Two years later a statistical branch of the commission was formed, and Balfour, now Deputy-Surgeon-General, was appointed to organize its Head-quarters. For the next fourteen years he compiled books on the health of the British Army. In 1867 he represented the English War Office at the International Statistics Congress at Florence, and twenty-two years later he appeared in the same capacity in Paris. In 1873 Balfour was promoted to Surgeon-General and put in charge of the hospital at Netley, then sent to Gibraltar a year later. In 1876 he retired on half-pay and henceforth was very active in various eminent societies, and was a corresponding member of the Academie Royale de Medicine de Belgique. In addition to all his statistical work, Balfour also carried out research in spirometry, meteorology and anthropology. He died at Wimbledon on 17 January 1891.

<u>Honours and Distinctions</u>

1859   Fellow of the Royal Society of London
1860   Fellow of the Royal College of Physicians of London
1867   Appointed Honorary Physician to the Queen
1888   President of the Royal Statistical Society

<u>Publications</u>

On the protection against smallpox afforded by vaccination (London, 1852)
Comparative Health of Seamen and Soldiers (London, 1871)
Inaugural address: Royal Statistical Society, 1888, 1889, 1890 (London, 1889-90)

<u>Biographical Studies</u>

<u>Edinb. Med. J.</u> <u>36</u>, (2), (Edinburgh, 1891), pp.778-82
Munk, W., Roll of the Royal College of Physicians, <u>4</u>, (London, 1878), pp.123-4
<u>The Times</u>, January 20, 1891, p.6
Boase, Frederick., Modern English Biography, <u>4</u>, supplement 1, (London, 1965), p.248

<u>Manuscript Locations</u>

L; LR; SaU

James BEGBIE (1798–1869)

Physician

Elected F.R.S.E. 19 February 1844

<u>Portraits, etc.</u>

Portrait: (Unknown), RCPE **

Born in Edinburgh in May 1798, little is known of James' early life. He received his early education at the local High School and then studied medicine at the University, where he graduated M.D. in 1821. He became the apprentice, and later assistant of Dr. Abercrombie, who was a considerable influence on his character and professional life. In 1850 he left Abercrombie to set up his own private consulting practice. In this capacity he was very popular and acquired a great reputation for his professional skills. Begbie was for many years Physician-in-Ordinary to the Queen in Scotland, and for thirty-seven years until his death, was physician to the Scottish Widows' Fund and Life Assurance Society. His major work consisted of an eight volume series of medical essays and memoirs published in Edinburgh. He died in Edinburgh on 26 August 1869 and was buried in the New Calton Burying-Ground.

<u>Honours and Distinctions</u>

1822   Fellow of the Royal College of Surgeons of Edinburgh
1847   Fellow of the Royal College of Physicians of Edinburgh
1848   Member of the Aesculapian Club
1850-52   President of the Medico-Chirurgical Society
1854-56   President of the Royal College of Physicians of Edinburgh
1854-56   Physician-in-Ordinary to the Queen in Scotland

<u>Publications</u>

De delirio trementi (Edinburgh, 1821)
On the utility of the actual cautery in some surgical diseases (Edinburgh, 1822)
On Erythema Nodosum, and its connection with the rheumatic diathesis (Edinburgh, 1850)
Contributions to practical medicine, 8 vols. (Edinburgh, 1862)

<u>Biographical Studies</u>

DNB., (London, 1968)
<u>Edinb. Med. J. 15,</u> (Edinburgh, 1869), pp.380-3
BMC., <u>13,</u> 1965, p.1205
Royal College of Physicians of Edinburgh. Historical sketch and laws from its institution to 1925, (Edinburgh 1925), p.11
Wemyss, H.L.W., A Record of the Edinburgh Harveian Society, 1782-1933, (Edinburgh, 1933), p.59
NUC., <u>43,</u> p.501

<u>Manuscript Locations</u>

E

James Warburton BEGBIE   (1826-1876)

Physician to the Royal Infirmary of Edinburgh

Elected F.R.S.E. 21 February 1870

<u>Portraits, etc.</u>

Photograph: see Comrie,J.D., History of Scottish
    Medicine, (London, 1932) **

Born on 19 November 1826, the second son of Dr. James Begbie, an Edinburgh physician, James was educated at the Grange Academy, Sunderland and the Edinburgh Academy. In 1843 he began to study medicine at Edinburgh University, graduating M.D. four years later. He then went to Paris to study under Cazenave and Devergie. In 1852 he returned to Edinburgh to become a family practitioner. Two years later he was appointed physician to the cholera hospital in Edinburgh but left that post the following year to take up the post of physician at the Royal Infirmary, where he remained for the next ten years. During that time he also gave lectures on the history of medicine and practice of physic. When his father died in 1869 he succeeded him as physician to the Scottish Widows' Fund Life Assurance Society. Begbie was extremely highly thought of as a consulting physician. In the profession he was a household name, and until his death he had the largest consulting physician's practice in Scotland. He was active in many Edinburgh medical societies, such as the Harveian Society and the Aesculapian Club. In 1873 he was appointed a member of the General Medical Council, a post he retained until his death. After a remarkably successful medical career, Begbie died in Edinburgh on 25 February 1876 from heart failure. He was buried in the Dean Cemetery. He was married to Anna Maria Reid, and had seven children by her.

<u>Honours and Distinctions</u>

1847-49	President of the Royal Medical Society, Edinburgh
1852	Fellow of the Royal College of Physicians of Edinburgh
1867-70	Honorary Librarian, Royal College of Physicians of Edinburgh
1875	Honorary LL.D.,Edinburgh University

<u>Publications</u>

Observations on the urine in cholera   (Edinburgh, 1849)
On the causes of death in the Scottish Widows' Fund Life Assurance Society from January 1853 to January 1860 (Edinburgh, 1860)
Handy book of medical information and advice (London, 1860)
The requirements for the proper study of medicine (Edinburgh, 1863)
Notice of some of the therapeutic effects of the bromide of potassium (Edinburgh, 1866)
Address in medicine; ancient and modern practice in medicine (Edinburgh, 1875)

<u>Biographical Studies</u>

Comrie, J.D., History of Scottish Medicine, <u>2</u>, (London, 1932), pp.616-17
<u>Proc. R. Soc. Edinb.</u>, <u>9</u>, (Edinburgh, 1878), p.209
Gray,J., History of the Royal Medical Society 1737-1937, (Edinburgh, 1952), pp.175-78, 203, 270-71, 321

<u>Manuscript Locations</u>

E; EU; L; SaU

Joseph BELL (1837-1911)

Surgeon

Elected F.R.S.E. 4 May 1874

<u>Portraits, etc.</u>

Photograph: see Comrie, J.D., History of Scottish
     Medicine, (London, 1932) **
Photograph: see MSS. Cat., EUL

Born in Edinburgh on 2 December 1837, the
eldest son of Benjamin Bell, surgeon, Joseph
was educated at the Edinburgh Academy and at
the University where he graduated M.D. in 1859.
He then worked as a hospital surgeon in the
Edinburgh Royal Infirmary being house-surgeon
and assistant to John Syme and house-physician
with John Gardiner, and a demonstrator of
anatomy under John Goodsir. From 1847 to 1863
Bell lectured on Botany in the Andersonian
School of Medicine in Glasgow and was lecturer of surgery at the extra-mural school of medicine
in 1864, 1879, and in 1889. He was an excellent systematic teacher, always attracting large
classes of students, but he was in his element as a clinical teacher, his theatre and wards
being constantly crowded with students. He was also a very good diagnostician, and his often
unexpected deductions were commemorated by one of his students, Sir Arthur Conan Doyle who bases
his world-famous detective Sherlock Holmes character on Bell. From 1873 he was for twenty-three
years editor of the 'Edinburgh Medical Journal'. Although not a profilic writer, his manual of
surgery was very successful, going through seven editions during his lifetime. Bell became the
first surgeon to the Sick Children's Hospital of Edinburgh in 1887. He died in Edinburgh on 4
October 1911.

<u>Honours and Distinctions</u>

1876   Secretary and Treasurer, Royal College of Surgeons of Edinburgh
1887   President, Royal College of Surgeons of Edinburgh
1897   President, Harveian Society, Edinburgh

<u>Publications</u>

A manual of the operations of surgery for the use of senior students, house-surgeons and junior
practitioners (Edinburgh and London, 1866)
Notes on surgery for nurses   (Edinburgh, 1887)
On the relation of the trained nurse to the profession and the public   (n.p., 1897)

<u>Biographical Studies</u>

Jaxby, J.M.E., Joseph Bell   (n.p., 1913)
Who's Who, 1898,   (London, 1898)
Comrie, J.D., History of Scottish Medicine, 2, (London, 1932), pp.454, 509, 510, 629, 675,
685, 712, 772
World Who's Who in Science, (Chicago, [1968]), p.147

<u>Manuscript Location</u>

E; EU; L

James BLACK (1787–1867)

Physician and Geologist

Elected F.R.S.E. 5 January 1857

<u>Portraits, etc.</u>

Engraving: see ref. in DSB, location unknown

Born in Scotland in 1787, nothing is known of James' early life. He trained as a surgeon and having been admitted Licentiate of the Royal College of Surgeons of Edinburgh in 1808, he joined the Royal Navy as an assistant surgeon the following year. He remained in this position during the Napoleonic Wars, and saw active service in the West Indies. He retired on half-pay and over the next few years, practised as a physician in Newton Stewart, Bolton and Manchester, until he retired to Edinburgh in 1856. During this period he decided to study medicine at Glasgow University, and duly graduated M.D. in 1820. Nine years later, he visited the United States, where he studied medical practice and education. After his return to Bolton, he made an extensive study of the town relating the physical and social enviroment to health and habits and noting industrial changes as well. Black was also a keen geologist and paleontologist, publishing several papers on these subjects. He was a founding member of the council of the Manchester Geological Society as well as a founder-member of the British Association in 1831. He lectured on geology at the Bolton Mechanics Institute and also served on a Bolton committee whose work led in 1853 to the establishment of one of the first municipal public libraries. Black died in Edinburgh on 30 April 1867.

<u>Honours and Distinctions</u>

1842   President of the British Medical Association
1860   Fellow of the Royal College of Physicians of London

<u>Publications</u>

A short inquiry into the capillary circulation of the blood (London, 1825)
A comparative view of the more intimate nature of fever (London, 1826)
A medico-topographical, geological and statistical sketch of Bolton and its vicinity (London, 1827)
A manual of the bowels and the treatment of their principal disorders from infancy to old age (London, 1840)
A memoir on the Roman Garrison at Mancunium, and its probable influence on the population and language of South Lancashire (Manchester 1849)

<u>Biographical Studies</u>

DNB., (London, 1968)
DSB., (New York, 1976)
Munk, W., Roll of the Royal College of Physicians of London, 1801-1825, <u>3</u>, (London 1878), p.277
<u>Br. med. J. 1</u>, (London, 1867), p.623
Sparke, A, Bibliographia Boltonensis (Manchester 1913), pp.26-27

<u>Manuscript Locations</u>

None known

Bruce Allan BREMNER   (1817-1895)

Physician

Elected F.R.S.E.   4 December 1875

<u>Portraits, etc.</u>

None known

Born in Dominica in 1817, Bruce Allan Bremner was the son of the Acting-Governor of the West Indies. He came to Scotland at an early age, where he went to school and eventually to the University of Aberdeen, graduating M.A. in 1837. His education in medicine began in Edinburgh where he took the degree of M.D. in 1840 and also the diploma of Licentiate of the Royal College of Surgeons of Edinburgh. After this he was engaged in practice in Bombay for thirteen years, returning home in 1853 to live in Perthshire for a time. Ten or so years later he settled at Streatham House, Canaan Lane, Edinburgh. Dr. Bremner was well-known for his connection with charitable organizations, such as the Association for Improving the Condition of the Poor, the Royal Edinburgh Hospital for Sick Children and the Edinburgh Medical Missionary Society. For a time Dr. Bremner was a director of the latter of these three. A large part of his life was devoted to philanthropic work. He died on 4 December 1895 and his obituary described him as a man of boundless sympathy and endless energy.

<u>Honours and Distinctions</u>

1844-55   Superintendent of Native Dispensary, Bombay
n.d.   Director of the National Securities Savings Bank
n.d.   Vice-President of the Indigent Old Men's Society

<u>Publications</u>

None known

<u>Biographical Studies</u>

<u>Trans. R. Scott. Soc. Arts, 7</u>, (Edinburgh 1868), p.201
List of graduates in medicine in the University of Edinburgh, from 1705 to 1866, (Edinburgh, 1867), p.121
The Medical Directory for 1875, (London, 1875), p.810; 1891, p.1194; 1896, p.1812
The Medical Register for 1893, (London 1893), p.207
<u>Br. med. J.</u>, 1895, ii, (London, 1895), p.1531

<u>Manuscript Locations</u>

None known

John BROWN   (1810-1882)

Essayist and Physician

Elected F.R.S.E. 17 January 1859

<u>Portraits, etc.</u>

Portrait: (Barbara Peddie), RCPE,
    see Comrie,J.D., History of Scottish Medicine,
    (London, 1932) **
Caricature: Kay, J.A., A series of portraits and
    caricature etchings, (Edinburgh, 1837)
Photograph: see MSS. Cat., EUL

Born at Biggar in Lanarkshire on 22 September 1810, son of the Rev. John Brown, the biblical scholar, John received his early education from his father. When the latter moved to Edinburgh in 1822 John entered William Steele's classical school where he remained for two years then he went to the High School. At the age of sixteen he went to Edinburgh University to study Arts but changed to medicine in 1828 and graduated M.D. in 1833 having served his apprenticeship with James Syme. Brown set up a practice in Edinburgh immediately after graduation, and remained in this position until his retirement through ill-health in 1876. Although a successful medical practitioner he was better known for his literary work. Most of this was closely connected with medicine although in 'Horae Subsecivae' he also dealt with such diverse subjects as poetry, art, nature, human character, the scenery of his country and the ways of dogs. To some of his contemporaries he was regarded as the 'Scottish Charles Lamb'. His most intimate friends were among the most eminent men of Edinburgh, including Lord Jeffrey and Lord Cockburn. His withdrawn nature and inclination towards melancholy meant, however, that he had to withdraw from society periodically. Brown died of pleurisy at 23 Rutland Street, Edinburgh on 11 May 1882.

<u>Honours and Distinctions</u>

1847   Fellow of the Royal College of Physicians of Edinburgh
1848   Librarian of the Royal College of Physicians of Edinburgh
1861   Honorary member of the Royal Medical Society
1874   Honorary LL.D., Edinburgh University

<u>Publications</u> *

Locke and Sydenham   (Edinburgh, 1850)
Horae Subsecivae, <u>1</u> (Edin., 1858); <u>2</u> (Edin., 1861); <u>3</u> (Edin., 1882)
Marjorie Fleming   (Edinburgh, 1862)
Jeems the doorkeeper, a lay sermon   (Edinburgh, 1864)
Sir Henry Raeburn and his works (Edinburgh, 1876) Printed for private collection
John Leech and other papers   (Edinburgh, 1882)

<u>Biographical Studies</u> *

MacLaren, E.T., Dr. John Brown and his sister Isabella   (Edinburgh, 1890)
Peddie, Alexander, Dr. John Brown, his life and work   (Edinburgh 1890)
Peddie, Alexander, Recollections of Dr. John Brown.... (London, 1893)
Clarkson, F.A., Dr. John Brown of Edinburgh   (Toronto, 1852)
DNB, (London, 1968)
<u>Edinb. Med. J.</u> <u>27</u>, (Edinburgh, 1882), pp.1130-37; <u>35</u>, (Edinburgh, 1890), pp.1048-62
Guthrie, D., <u>Janus in the Doorway</u>,   (London, 1963), pp.116-129

<u>Manuscript Locations</u>

E; EPH; EU; L; LR; O; SaU

William Alexander Francis BROWNE   (1805–1885)

Alienist

Elected F.R.S.E. 1861

<u>Portraits, etc.</u>

Photograph: see Carmont, J.H., The Crichton Royal
    Institution, Dumfries, (Leicester, 1896) **

Browne was born near Stirling in 1805, the son
of an officer in the Cameronian Regiment, who
died in the Peninsular War. He was educated at
Stirling High School, and then studied medicine
at the University of Edinburgh, graduating M.D.
in 1826. He was President of the Medical
Society there, and as well as assisting George
Combe in his phrenological researches, he made
contributions to the 'Phrenological Journal'.
Soon after the commencement of his profession
in a practice in Stirling, Dr. Browne was
appointed Medical Superintendent to the Hospital for the Insane at Montrose, and carried out his
duties there with skill and high credit. In 1839 Dr. Browne accepted the position of Medical
Superintendent to the New Crichton Asylum at Dumfries. He retained this position for eighteen
years, improving and developing the Asylum. In 1857 he accepted the role of First Commissioner
in Lunacy for Scotland, with Sir James Coxe as his colleague. In 1870 his eyesight was damaged,
and after an unsuccessful operation he totally lost his sight. He retained his intellectual
vision, having books read to him, and he himself was not an infrequent contributor to periodical
literature. He died in Dumfries on 2 March 1885, leaving four children, one of whom, the eldest
Dr. Crichton-Browne followed in his footsteps.

<u>Honours and Distinctions</u>

1857 First Commissioner in Lunacy for Scotland

<u>Publications</u> *

What Asylums were, are, and ought to be   (Edinburgh, 1835)
Cottage asylums   (London, 1861)
Endemic degeneration   (London, 1862)
Epileptics: their mental condition. A lecture   (London, 1865)

<u>Biographical Studies</u> *

<u>Br. Med. J.</u>, 1885, (London, 1885), p.568
Boase, F., Modern English Biography, 1, (London, 1965), p.446
Gray, J., History of the Royal Medical Society, (Edinburgh, 1952), pp.221-222,   319
Biographisches Lexikon der Hervorragenden Aerzte, 1, (Berlin, 1929), p.725
Index Catalogue of the Library of the Surgeon-General's Office, U.S. Army, 2nd series
(Washington, 1881), p.494
NUC, 79, p.636

<u>Manuscript Locations</u>

None known

John CHIENE (1843-1923)

Professor of Surgery at the University of
Edinburgh

Elected F.R.S.E. 4 May 1874

Portraits, etc.

Photograph: see Comrie, J.D., History of Scottish
    Medicine, (London, 1932) **

Born in Edinburgh on 25 February 1843, the son
of George Todd Chiene, a chartered accountant,
John was educated at Mr. Hunter's school and at
the Edinburgh Academy. He began his medical
studies at Edinburgh University. In 1860 he
travelled to Paris, Berlin and Vienna to study
further before graduating with honours from
Edinburgh five years later. Soon after James
Syme made him his house surgeon for a year, and
from 1866 to 1870 he was a demonstrator in
anatomy with John Goodsir and Professor William Turner respectively. He then became a lecturer
on surgery in the extra-mural school. A year later he was assistant-house surgeon to the Royal
Infirmary, becoming junior surgeon seven years later. In 1882 he was appointed Professor of
Surgery at Edinburgh University and remained in this position until 1909. Chiene had worked with
Joseph Lister, and was very active in promoting the latter's ideas on antiseptic surgery at home
and abroad. He established the first teaching laboratory for bacteriology in the United Kingdom
and was a pioneer of the Edinburgh Ambulance Service. He was considered to be a great teacher,
his outspokenness earned him the nickname 'Honest John' among his students, who treasured such
sayings of his as, 'a pimple on a man's nose is of more interest to him than a sarcoma in his
neighbour's thigh'. In his earlier days he had been a keen player of rugby and was the first
president of the Scottish Rugby Union. During the Boer War he was consulting surgeon to H.M.'s
Forces in South Africa. Chiene received many honours, including being made an Honorary Fellow of
the American, Finnish and Danish Surgical Associations. He died in Edinburgh on 29 May 1923.

Honours and Distinctions

1865   President of the Royal Medical Society of Edinburgh
1897   President of the Royal College of Surgeons of Edinburgh
1900   Companion of the Bath
1903   President of the Harveian Society, Edinburgh
1908   Hon. D.Sc., Sheffield

Publications *

Contributions to Surgical and General Pathology (Edinburgh, 1870)
Lectures on Surgical Anatomy (Edinburgh, 1878)
Lectures on the elements or first principles of surgery (Edinburgh, 1882)
Germs and the Spray (Edinburgh, 1884)
Lectures on Surgery (Edinburgh, 1898-9)

Biographical Studies

Comrie, J.D., History of Scottish Medicine, 2, (London, 1932), p.673
Who's Who, London, 1898
Who's Who, London, 1923
Edinb. Med. J. 30, (Edinburgh, 1923), pp.285-8

Manuscript Locations

E; ES; EU

Thomas Beath CHRISTIE (-1892)

Surgeon

Elected F.R.S.E.   15 April 1872

<u>Portraits, etc.</u>

None known

Nothing is known of Thomas Christie's early life. He graduated M.D. from St. Andrews University in 1854 and soon afterwards was appointed assistant medical officer at Pembroke House, Hackney, an asylum instituted by the East India Company for mentally ill soldiers of their army. In 1867 he moved to Clifton near York to become medical superintendent of the North Riding Asylum. He only remained here for three years, however, before moving to the Royal India Asylum, Ealing in 1870, to take up a similar post. From 1873 he also assumed the duties of Surgeon-Major of the 5th Battalion of the Rifle Brigade. Christie retired in 1891, and died shortly after, on 15 January 1892.

<u>Honours and Distinctions</u>

1858   Fellow of the Royal College of Physicians of Edinburgh
1873   Fellow of the Royal College of Physicians of London
1877   Companion of the Order of the Indian Empire

<u>Publications</u>

None known

<u>Biographical Studies</u>

Munk, W., Roll of the Royal College of Physicians of London, <u>4</u>, (London 1925), p.210
<u>Transactions of the Medico-Chirurgical Society of London</u>, <u>75</u>, (London, 1892), p.29
<u>Br. med. J.</u> <u>1</u>, (London, 1892), p.312
Crawford, D.G., History of the Indian Medical Service, <u>1</u>, (London, 1914), p.172

<u>Manuscript Locations</u>

None known

James Scarth COMBE (1796-1883)

Surgeon

Elected F.R.S.E.   16 December 1850

<u>Portraits, etc.</u>

None known

Born in Leith, Edinburgh on 5 January 1796,
nothing is known of James' early life. He
obtained his medical diploma in 1814 and was
precluded from a career in Europe as a military
surgeon by the ending of the Napoleonic Wars.
Before he had completed his studies he went to
India for a short period, then returned to
Scotland to set up practice as a surgeon. Later
he moved to Edinburgh and for many years acted
as an examiner and assessor. Combe primarily
distinguished himself with his paper on
anaemia, read before the Medico-Chirurgical
Society of Edinburgh in 1822. He was the first
to describe pernicious anaemia, and suggested
that this condition was primarily one of the
digestive system. His description preceded
Addison's by almost thirty years. He was
married and had one daughter and two sons. He
died on 14 February 1883.

<u>Honours and Distinctions</u>

1823   Fellow of the Royal College of Surgeons of Edinburgh
1851   President of the Royal College of Surgeons of Edinburgh

<u>Publications</u>

"History of a case of anaemia", <u>Trans. med.-chir. Soc. Edinb.</u>, <u>1</u>, (Edinburgh, 1824), pp.194-204
On the poisonous effects of the Mussel (Mytius edulis), <u>Edinb. Med. Surg. J.</u>, <u>29</u>,
(Edinburgh, 1828), pp.86-96

<u>Biographical Studies</u>

<u>Edinb. Med. J.</u>, <u>28</u>, (2) (Edinburgh, 1883), pp.862-3
Comrie, J.D., History of Scottish Medicine, <u>2</u>, (London, 1932), p.492

<u>Manuscript Locations</u>

E

Sir John Rose CORMACK   (1815–1882)

Physician to the Royal Infirmary of Edinburgh
and the Fever Hospital

Elected F.R.S.E. 6 February 1843

<u>Portraits, etc.</u>

Photograph: see <u>Edinburgh Medical Journal</u>, <u>17</u>,
   (Edinburgh, 1905) **

Born at Stow, Midlothian, on 1 March 1815, the
son of the Rev. John Cormack, minister of the
parish, John studied first at the local parish
school, then at the Edinburgh High School. He
then went on to study medicine at Edinburgh
University, graduating M.D. in 1837. He won a
gold medal for his inaugural thesis, "On the
presence of air in the organs of circulation",
the year previously having won the Harveian
Prize. After he had made a tour of Italy and
Spain, and visited Paris, Cormack returned to Edinburgh to become physician at the Royal
Infirmary. He founded the 'Edinburgh Monthly Journal of Medical Science' in 1841, and was its
editor for the next six years. He then moved to London, where he established the 'London Medical
Journal of Medicine' and edited 'The Associated Medical Journal' from 1853–55, which was the
forerunner of the 'British Medical Journal'. On account of ill-health he was forced to settle
in Orleans and became, in 1869, Principal English Physician at the British Embassy in Paris. In
the following year he took the Paris degree of M.D. His hopes of a peaceful life were shattered
when the Franco-Prussian War broke out, but he stayed in Paris. Throughout the civil war and
the siege of Paris he was of assistance to the wounded of both sides, for which he was awarded a
knighthood by Queen Victoria, and enrolled as a Chevalier of the Legion of Honour by the French.
He was appointed physician to the Hertford British Hospital, and in his later years became a
leader of his British colleagues in Paris. He died at his home in the Rue St. Honore, Paris, on
13 May 1882.

<u>Honours and Distinctions</u> *

1841   Fellow of the Royal College of Physicians of Edinburgh
1872   Fellow of the Royal College of Surgeons of England
1872   Knighted
1872   Fellow of the Royal College of Physicians of London

<u>Publications</u> *

A Treatise on the Chemical, Medicinal and Physiological Properties of
Creosote .... being the Harveian Prize Dissertation for 1836 (Edinburgh, 1836)
Prize thesis: Inaugural Dissertation "On the Presence of Air in the Organs
of Circulation"   (Edinburgh, 1837)
Natural History, Pathology and Treatment of the Epidemic Fever, at present
prevailing in Edinburgh and other towns   (Edinburgh, 1843)

<u>Biographical Studies</u> *

DNB., (London, 1968)
<u>Proc. R. Soc. Edinb.</u>, <u>14</u>, (Edinburgh, 1887), p.561
Munk, W., The Roll of the Royal College of Physicians of London, <u>4</u>, (London, 1955), p.200
<u>Edinb. Med. J.</u>, <u>27</u>, (Edinburgh, 1882), pp.1137–40
Royal College of Physicians of London, Library Catalogue, (London 1912), pp.288–9

<u>Manuscript Locations</u>

E; EU

Sir James COXE   (1811-1878)

Commissioner in Lunacy for Scotland

Elected F.R.S.E.   4 December 1854

<u>Portraits, etc.</u>

None known

Born at Gorgie, Midlothian in 1811, fourth son of Robert Coxe, little is known of Sir James' early life. His father died when he was young, and his mother was left to bring up the family. He studied medicine at Gottingen, Heidelberg and Paris before returning to Edinburgh to study at the University. He duly graduated M.D. in 1835, and soon afterwards joined the Royal College of Physicians of Edinburgh. He was interested in physiology and pathology of the mind, and was contemplating setting up a private asylum in Edinburgh, when in 1855 he was appointed to a Royal Commission for the management of the insane in Scotland. Sir James was responsible for writing the report of the commission which disclosed chaotic arrangements, neglect and cruelty in the care of the insane to such an extent that when presented to Parliament in 1857, members were sufficiently shocked to pass Lord Moncrieff's Lunacy Act that same year. As a result of this report Sir James was also made a paid commissioner, a post he held until his death. He wrote the first fifteen reports of the commissioners, and visited English and Continental establishments for the care of the mentally ill, making himself an expert in their methods, and a promoter on improvements in provisions for the mentally ill at home. At his suggestion, amendments were thus made to the Lunacy Act in 1857, 1862 and 1866. In later years he devoted much of his spare time to supervising the memoirs of his uncle George Combe. He died on 9 May 1878 at Folkestone, after returning from a holiday in Paris.

<u>Honours and Distinctions</u>

1837   Fellow of the Royal College of Physicians of Edinburgh
1863   Knighted
1872   President of the Psychological Association

<u>Publications</u>

Lunacy in its relation to the State (London, 1878)
Editor of Andrew Combe's The Physiology of Digestion (Edinburgh, 1849) and the Principles of Physiology (Edinburgh, 1852)
Revised with Robert Coxe, George Combe's, The Constitution of Man considered in relation to external objects (Edinburgh, 1860)

<u>Biographical Studies</u>

<u>Proc. R. Soc. Edinb.</u>, <u>10</u>, (Edinburgh, 1880), pp.15-17
Boase, F., Modern English Biography, <u>1</u>, (London, 1965), p.743

<u>Manuscript Locations</u>

E; L

William CRAIG   (1832-1922)

Surgeon and Lecturer

Elected F.R.S.E. 4 January 1875

<u>Portraits, etc.</u>

Photograph: see <u>Edinburgh Medical Journal, 28,</u>
     (Edinburgh, 1922) **

Born at High Ploughland, Avondale, on 28 March
1832, the son of John Craig, a farmer of High
Ploughland, William received his early
education at the local parish school. He then
went to Glasgow University to study Arts and
Divinity with a view to becoming a minister,
but later moved to Edinburgh University and
graduated M.B. and C.M. there in 1868. Two
years later he graduated M.D. He began his
professional life as an instructor in practical
pharmacy at the New Town Dispensary, and soon
became a lecturer on materia medica and therapeutics in the extra-mural medical school at Minto
House and Surgeons' Hall. His interests turned increasingly towards teaching and examining
branches of his profession, and he was for many years examiner in the above subjects for the
Public Health and Dental Diploma qualification of the Royal Colleges, as well as for the
Edinburgh University medical degree course. Craig was very interested in the natural sciences
and botany, and, in this connection, was active in the local Botanical Society. He was also
active in Edinburgh's medical life, being the editor of the 'Transactions of the Medico-
Chirurgical Society' from 1882-1909, Fellow and Treasurer of the Obstetrical Society for thirty-
five years from 1875 and member of the President's Council of the Royal College of Surgeons. He
was also a lecturer in the Edinburgh Medical College for Women from its inception. He was one of
the oldest members of the Scottish Alpine Botanical Club, a Director of the Edinburgh Dental
Hospital and a Manager of the Edinburgh Savings Bank. In his later years he was the oldest
living medical practitioner in Edinburgh. He died at his home at 71 Bruntsfield Place,
Edinburgh, on 3 February 1922.

<u>Honours and Distinctions</u>

1874   Fellow of the Royal College of Surgeons of Edinburgh
1887   President of the Edinburgh Botanical Society
1900   Honorary Secretary of the Edinburgh Botanical Society

<u>Publications</u>
On the influence of variations of electric tension as the remote cause of epidemic and other
diseases (London 1859)
Manual of materia medica and therapeutics. 4th edition (Edinburgh, 1887)
Posological tables; appendix on poisons; index of diseases and medicines (Edinburgh, 1899)
Craig's posological tables   (Edinburgh, 1933)

<u>Biographical Studies</u>

<u>Edinb. Med. J., 28,</u> (Edinburgh 1922), pp.59, 179-80
<u>Proc. R. Soc. Edinb., 42,</u> (Edinburgh, 1922), pp.390-1
Eddington,A., Edinburgh and the Lothians at the opening of the twentieth century
(Edinburgh, 1904), p.219

<u>Manuscript Locations</u>

E

Sir James CRICHTON-BROWNE   (1840-1938)

Physician and Pyschologist

Elected F.R.S.E.   3 January 1870

<u>Portraits, etc.</u>

Woodcut: (Unknown), NPG **
Photograph: (W. Stoneman, 1917), NPG

Born in Edinburgh on 29 November 1840, son of
Dr W.A.F. Browne, H.M's Commissioner in Lunacy
for Scotland, James was educated at Dumfries
Academy and Trinity College, Glenalmond, in
Perthshire. He then studied medicine at
Edinburgh University, graduating with honours
in 1862, his thesis "On Hallucinations", being
highly commended. The same year found him as a
member of the Royal College of Surgeons. He
spent a year in Paris and London doing further
studies. In 1865 he was appointed Medical
Superintendent of the Borough Asylum, Newcastle-on-Tyne. A year later he was promoted to
Medical Superintendent of the lunatic asylum for the West Riding of Yorkshire, at Wakefield, a
post which he held for ten years. The asylum came to be considered the leading school of medical
psychology in Europe. Under Crichton Browne's superintendency, the medical staff increased, a
pathological museum was formed, and a physiological laboratory instituted, in which Ferrier's
first series of experiments into the functions of the brain were conducted. Crichton Browne was
one of the founders and the editor of the magazine 'Brain'. In 1879 an honorary degree of LL.D.
from St. Andrew's University was conferred on him. As a result of his essay "Overpressure in
Schools", he was asked to consider that subject in relation to elementary schools in London. His
report was published as a Parliamentary Paper. An important aspect of his work was the care of
children of the poor, and his efforts to ameliorate the conditions of those children were
recognised by his placement on the London School Board. In 1883 he was made a Fellow of the
Royal Society of London, and between 1905 and 1913 he was President of the London Dumfriesshire
Association. Browne died in 1938.

<u>Honours and Distinctions</u> *

1879   Honorary LL.D., St. Andrews University
1883   Fellow of the Royal Society of London
1886   Knighted

<u>Publications</u> *

The prevention of senility and a sanitary outlook   (London, 1905)
The story of the brain   (Edinburgh and London, 1924)
Physical effiency in children   (London, 1902)
Physiology and psychology of educational methods and work   (n.p., 1907)
The doctor's second thoughts   (London, 1931)
The doctor's after-thoughts   (London, 1932)

<u>Biographical Studies</u> *

Who's Who 1889, (London, 1889)
Who's Who 1898, (London, 1898)
Royal College of Physicians, Library Catalogue, <u>1</u>, (Edinburgh, 1898), p.208
The Medical Register for 1893, (London, 1893), p.224
BMC., <u>28</u>, 1965, pp.2, 3, 4

<u>Manuscript Locations</u>

E; EU

Robert James Blair CUNYNGHAME   (1841-1903)

M.D. of Cronan, Superintendent of Statistics

Elected F.R.S.E.   6 February 1871

<u>Portraits, etc.</u>

None known

Born in January 1841, little is known of Cunynghame's early life. He studied medicine at Edinburgh University, graduating M.D. in 1862 and he received the gold medal in the Forensic Medicine Class. After a spell as a resident physician in the Royal Infirmary, he pursued his medical studies in London, Berlin, Paris and Vienna. He then went into medical service in the army, serving four years in India. On his retirement he settled in Edinburgh, where he took his Fellowship of the Royal College of Surgeons of Edinburgh in 1866, and membership of the Royal College of Physicians of Edinburgh in 1877. He was an expert physiologist, and was an excellent teacher of that subject. He also gave classes on pathology, and held the post of conservator of the Museum of the Royal College of Surgeons of Edinburgh. for many years. Cunynghame was a physician at the Royal Hospital for Sick Children and for many years was on the Board of Directors, remaining in that position until his death. All his working life he had a private general practice, which he gave up when he was appointed Superintendent of Statistics in the Register General's Department for Scotland. He was president of several societies; served as surgeon in the Queen's Edinburgh Volunteer Brigade and acted on the board of management of charitable and philanthropic institutions. Cunynghame died on 23 December 1903 at Rothesay Place, Edinburgh.

<u>Honours and Distinctions</u> *

1862   President of the Royal Medical Society of Edinburgh
1866   Fellow of the Royal College of Surgeons of Edinburgh
1879   Member of the Aesculapian Club
1887   Secretary and Treasurer of the Royal College of Surgeons of Edinburgh
1891   President of the Royal College of Surgeons of Edinburgh

<u>Publications</u>

'Case of suppression of urine', <u>Edinb. Med. J.</u> (Edinburgh, 1874)

<u>Biographical Studies</u> *

Gray, J., History of the Royal Medical Society 1737-1937, (Edinburgh, 1952), pp.145,235,322
<u>Br. med. J.</u> (London, 1904), pp.52-53
<u>The Lancet,</u> (London, 1904), p.201
<u>Trans. R. Soc. Edinb.,</u> 36, (Edinburgh 1892), p.738
The Medical Directory for 1905, (London, 1905), p.1807
The Medical Register for 1893, (London 1893), p.344

<u>Manuscript Locations</u>

None known

James DUNCAN   (1810–1866)

Senior Surgeon to the Royal Infirmary

Elected F.R.S.E.   7 December 1857

<u>Portraits, etc.</u>

None known

Born in Perth on 2 November 1810, the only son of Mr. Duncan of Duncan, Flockhart and Company, James received his early education first at Perth then at the High School in Edinburgh. He studied medicine at Edinburgh University and graduated M.D. in 1834 with the thesis 'On Empyema'. After studying anatomy and surgery at medical schools in France, Germany, Austria and Italy for two years he returned to Edinburgh to act as house-surgeon, first to James Gregory and then to Robert Liston. He also spent some time with Liston at the hospital of University College in London. By 1837, he had set himself up as a medical practitioner in Edinburgh and in due course enjoyed great eminence. He was senior surgeon to the Royal Infirmary for many years and in that capacity gave lectures on clinical surgery. He contributed several papers to medical journals and in his leisure hours, pursued his favourite hobby of landscape painting. He also acted as medical officer for the Scottish Provident Institution and was consulting surgeon to the New Town Dispensary. In July 1866 he went on a tour of France with his wife and family and after a cholera attack died in Tours on 16 August 1866.

<u>Honours and Distinctions</u>

1834   Fellow of the Royal College of Surgeons of England
1835   Fellow of the Royal College of Surgeons of Edinburgh

<u>Publications</u> *

On foreign bodies in the air passage (Edinburgh, 1835)
Neuralgia of stump after amputation; secondary amputation; return of the neuralgia (Edinburgh, 1844)
Case of fibrous tumour surrounding the right sciatic nerve (Edinburgh, 1844)
Surgical cases: strangulated femoral hernia; gangrene; death (Edinburgh, 1844)

<u>Biographical Studies</u>

<u>Proc. R. Soc. Edinb.</u>, <u>6</u>, (1866–69), (Edinburgh, 1869), pp.18–19
Plarr, V.G., Lives of the Fellows of the Royal College of Surgeons of England, <u>1</u>, (Bristol and London 1930), p.356

<u>Manuscript Locations</u>

E; SaU

James Matthews DUNCAN (1826-1890)

Physician-Accoucheur at St.Bartholomew's
Hospital, London

Elected F.R.S.E. 16 February 1863

<u>Portraits, etc.</u>

Portrait: (H. Kerr), RCPE
Photograph: see Comrie,J.D., History of Scottish
    Medicine, (London, 1932) **
Portrait: (Unknown), NPG

Born in Aberdeen in April 1826, the fifth child
of William Duncan, a commissioner and shipping
agent, James received his early education at
the local grammar school. After obtaining his
M.A. degree from Marischal College, Aberdeen in
1843, he went on to study medicine at Aberdeen
and Edinburgh and three years later graduated
M.D. from Marischal College. Following a brief
visit to Paris, he settled in Edinburgh and became an assistant to James Young Simpson. On 4
November 1847 he took part in the historic inhalation of chloroform along with James Young
Simpson and Alexander Keith. He then went on a continental tour with the Marquis of Bute and on
his return to Edinburgh set up practice as an obstetrician. In 1853 he began to teach midwifery
as an extra-mural lecturer and eight years later was appointed physician and consultant to the
Royal Infirmary. He was instrumental in the founding of the Royal Hospital for Sick Children in
Edinburgh. As well as having a successful practice he also published several works including the
massive but invaluable 'Fecundity, fertility, and sterility' published in 1866. Four years later
he accepted the post of Physician Accoucheur and Lecturer on Midwifery at St.Bartholomew's
Hospital in London. Here he soon established a large practice and his 'Clinical Lectures on
Diseases of Women', published in 1879, became a standard work. He was highly regarded as a
teacher and examiner for several universities. He was married to Jane Hart Hotchkis by whom he
had nine children. He died in Baden-Baden on 1 July 1890.

<u>Honours</u> and <u>Distinctions</u>

1851   Fellow of the Royal College of Physicians of Edinburgh
1862   Honorary Librarian of the Royal College of Physicians of Edinburgh
1875   Honorary LL.D., Edinburgh University
1882   Fellow of the Royal College of Physicians of London
1883   Fellow of the Royal Society
1883   President of the Obstetrical Society of London
1883   Crown Member of the General Medical Council

<u>Publications</u>

Fecundity, fertility and sterility, (Edinburgh, 1866)
Researches in Obstetrics, (Edinburgh, 1868)
Clinical Lectures on Diseases of Women, (London, 1879)

<u>Biographical</u> <u>Studies</u>

Munk,W., Lives of the Fellows of the Royal College of Physicians of London, <u>4</u>, (London, 1878),
pp.286-7
Boase, F., Modern English Biography, <u>5</u>, (London, 1965), pp.168-9
<u>Edinb. Med. Surg. J.</u>, <u>36</u>, (1), (1890), pp.392-7, and <u>54</u>, (1947), pp.593-609

<u>Manuscript</u> <u>Locations</u>

EPH; EU; L

John DUNCAN (1839-1899)

Surgeon

Elected F.R.S.E.   3 January 1878

<u>Portraits, etc.</u>

Photograph: Medical Archives Centre, Royal
    Infirmary of Edinburgh **

Born on 18 August 1839, the son of James
Duncan, M.D., a physician of Edinburgh, John
received his early education at the city's High
School. He then went to Edinburgh University to
study Arts but later decided to take up
medicine and duly graduated M.D. with
distinction in 1862. He then became house
surgeon to James Syme but gave this up eighteen
months later in order to study abroad. He
returned two years later and was appointed
assistant surgeon at the Royal Infirmary where
he became closely associated with Joseph Lister. Shortly afterwards he began to lecture in
surgery at the extra-mural school and in time became one of the most accomplished scientific
teachers of clinical surgery. He is also reputed to have demonstrated that in chlorosis the
haemoglobin content is lessened but the number of red cells remains almost unchanged. In 1887 he
was appointed senior surgeon to the Royal Infirmary. After twenty years he left the Infirmary in
1895 and spent his last years as a Commander in the Bombay Army. He was a well-known figure in
Edinburgh because of his tall silk hat, long grey beard and pair of high-stepping horses which
he drove in a yellow "dog cart". He died on 24 August 1899.

<u>Honours and Distinctions</u>

1864   Fellow of the Royal College of Surgeons of Edinburgh
1889   President of the Royal College of Surgeons of Edinburgh

<u>Publications</u>

On the galvano-puncture of aneurisms (Edinburgh, 1866)
The treatment of aneurism by electrolysis. With an account of an investigation into the action
of galvanism on blood and on albuminous fluids (Edinburgh, 1867)
On galvano-puncture of naevus (Edinburgh, 1870)
On the surgical applications of electricity (Edinburgh, 1873)
The treatment of naevus (Edinburgh, 1876)

<u>Biographical Studies</u>

Hodson, J., ed., Angroma and other papers, with memoir by J. Chiene (Edinburgh, 1900)
<u>Edinb. Med. J., 6,</u> (Edinburgh, 1899), pp.389-90
<u>Proc. R. Soc. Edinb., 23,</u> (Edinburgh, 1901), p.8

<u>Manuscript Locations</u>

None known

James DUNSMURE   (1814-1886)

Surgeon

Elected F.R.S.E.   19 February 1866

<u>Portraits, etc.</u>

None known

Born on 23 April 1814, the third son of James Dunsmure, secretary of the Fisheries Board of Scotland, James received his early education at the High School in Edinburgh. He then studied medicine at Edinburgh University, and graduated M.D. in 1835 with the thesis 'On Scarlet Fever'. After serving an apprenticeship with Robert Liston, he began practising as a surgeon, being appointed senior acting surgeon in 1854. In 1856 he started lectures on clinical surgery at the extra-mural school in Edinburgh. He was also assistant to John Lizars for a time. Later he became consulting surgeon at the Royal Infirmary, and in 1886 he was again a lecturer in the extra-mural school, this time on the diseases of children. Dunsmure was active in many local medical societies, including the Harveian and the Aesculapian Clubs. He was also an elder at St. Andrew's Church of Scotland in Edinburgh for forty years. He died at his home in Queen Street on 15 October 1886.

<u>Honours and Distinctions</u>

1841   Fellow of the Royal College of Surgeons of Edinburgh
1865   President of the Royal College of Surgeons of Edinburgh
1869   President of the Harveian Society, Edinburgh
n.d.   Member of the Royal Company of Archers

<u>Publications</u>

A probationary essay on varix and the treatment by compression as recommended by Velpeau (Edinburgh, 1841)

<u>Biographical Studies</u>

Comrie, J.D., History of Scottish Medicine, 2, (London, 1932), pp.629, 714
<u>Edinb. Med. J.</u>, 32, (1886), pp.477-9

<u>Manuscript Locations</u>

EU

Robert DYCE   (1798–1869)

Professor of Midwifery at the University of
Aberdeen

Elected F.R.S.E.   7 March 1864

<u>Portraits, etc.</u>

Portrait: (Unknown), Marischal College, Aberdeen
    see Comrie, J.D., History of Scottish Medicine,
    (London, 1932) **

Born in Aberdeen in November 1798, the eldest
son of Dr. William Dyce, an eminent physician
of the city, Robert studied Arts at Marischal
College, Aberdeen, graduating M.A. in 1816. He
then took up medicine, studying at Aberdeen
University, Edinburgh and London before joining
the army. He served at the Military Hospital at
Chatham before taking a staff appointment in
Mauritius in 1821. He stayed there for five
years, becoming very popular with the English residents, possibly because he refused to accept
any fees for medical attendance. This was compensated by the valuable gifts given to him by
grateful patients. Later he was transferred to the Cape, where he remained for several years. In
both places he made extensive studies of the local natural history. He returned to Aberdeen in
1833, but left soon after, to take up a staff appointment at Maidstone. Three years later, on
the death of his father, he took over the latter's extensive practice in Aberdeen. Partly as a
result of his popularity in this capacity, he was appointed lecturer of Midwifery at Marischal
College in 1841, and when the two colleges became one university, he was appointed Professor of
Midwifery. He was highly thought of as a teacher and as a practising physician, and for the
medical assistance he gave to the poor. He died in Edinburgh on 11 January 1869.

<u>Honours and Distinctions</u>

None known

<u>Publications</u>

On the importance of the pulse in relation to chloroform   (London, 1857)
On the identity of Morrhua punctata and Morrhua vulgaris (London, 1860)
Case of caesarean section (Edinburgh, 1862)

<u>Biographical Studies</u>

<u>Proc. R. Soc. Edinb.</u>, <u>7</u>, (Edinburgh, 1872), pp.9–10
Comrie, J.D. History of Scottish Medicine, <u>2</u>, (London, 1932), pp.564–5
Biographisches Lexikon der Hervorragenden Aerzte, <u>2</u>, (Berlin, 1929), p.363

<u>Manuscript Locations</u>

None known

Alexander Mackenzie EDWARDS (-1868)

Lecturer on Surgery

Elected F.R.S.E.   20 January 1862

<u>Portraits, etc.</u>

None known

Little is known of Edwards' early life or education. In 1855, Edwards was elected a Fellow of the Royal College of Surgeons of Edinburgh. He gave lectures on surgery at the Extra-academical School of Medicine and Surgery in 1859. He acted as a demonstrator of anatomy in King's College, London, and at Edinburgh University. He was appointed a surgeon in the New Town Dispensary, followed by a surgeon's post with the Midlothian Coast 5th. Artillery. Edwards was a member of various societies, including the Pathological Society, London, the Medico-Chirurgical Society and the Royal Physiological Society, based in Edinburgh.   He was the author of many publications, including, 'Hints on rupture and trusses, and remarks on instruments for the relief of deformity; being a guide to those who require the aid of the surgical machinist', (1847). Edwards died in 1868.

<u>Honours</u> <u>and</u> <u>Distinctions</u>

1852   Member of the Royal College of Surgeons of England
1855   Fellow of the Royal College of Surgeons of Edinburgh
1855   Member of the Royal Physical Society, Edinburgh
n.d.   Member of the Pathological Society, London
n.d.   Member of the Royal Physiological Society, Edinburgh

<u>Publications</u>

Hints on rupture and trusses, and remarks on instruments for the relief of deformity; being a guide to those who require the aid of the surgical machinist (Edinburgh, 1847)
Evidence in favour of .... candidate for the appointment of assistant-surgeon to the Royal Infirmary (Edinburgh, 1856)
Agreement of a course of lectures on the principles and practice of surgery (Edinburgh, 1858)
The Ambulance Surgeon; or, Practical observations on gunshot wounds; co-editor T.W. Nunn (Edinburgh, 1862)
Clinical cases (Edinburgh, 1863)

<u>Biographical</u> <u>Studies</u>

Comrie, J.D., History of Scottish Medicine, <u>2</u>, (London, 1932), p.629
<u>Proc.</u> <u>R.</u> <u>Soc.</u> <u>Edinb.</u>, <u>5</u>, (Edinburgh, 1866), p.467
BMC., <u>59</u>, 1960, p.607
Royal College of Physicians, Library Catalogue, <u>1</u>, (Edinburgh, 1898), p.435

<u>Manuscript</u> <u>Locations</u>

None known

Sir Joseph FAYRER, Bart   (1824-1907)

Physician Extra-ordinary to the King

Elected F.R.S.E.   3 January 1859

<u>Portraits, etc.</u>

Portrait: (Unknown), NPG **

Born in Plymouth on 6 December 1824, the second son of Commander Robert J. Fayrer, Royal Navy, of Milnthorpe, Westmorland, a pioneer of ocean steam navigation, Joseph received his early education privately, at Dalrymple in Ayrshire, and at a Liverpool school. At sixteen he sailed to the West Indies as a midshipman, and three years later accompanied his father on a journey to Bermuda. Here he became interested in the study of medicine through his experience of yellow fever, and on his return to Britain in 1844, he began to study medicine at Charing Cross Hospital, London. He graduated M.D. three years later, and immediately joined the Naval Medical Service. He soon resigned, however, to accompany Lord Mount-Edgcumbe on his continental travels. At Palermo, Fayrer had his first experience of treating war wounds, and in Rome he graduated M.D. in 1849. A year later he was gazetted to the Bengal Medical Service, and his services in India earned him a steady rise through the ranks until he was promoted to the brevet rank of surgeon for his services in the Mutiny. In 1859 he obtained his M.D. from Edinburgh University. At that time he was Professor of Surgery at Calcutta. Four years later he was elected President of the Faculty of Medicine and three years after that, of the Asiatic Society. In 1869 he was made Surgeon to the Viceroy, and accompanied the Duke of Edinburgh to North-West India a year later. From 1872 until he finally retired as Surgeon-General in 1895, he was President of the India Office Medical Board. He still kept busy, however, with his private practice, committee work and writing. He spent his later years at Falmouth in active retirement, where he died on 21 May 1907.

<u>Honours and Distinctions</u>

1872-95   President of the Medical Board, India Office
1875   Knight Commander of the Order of the Star of India
1896   Created Baronet
1901   Physician Extraordinary to the King

<u>Publications</u> *

Clinical surgery in India   (London, 1866)
The thanatophidia of India (London, 1872)
European child-life in Bengal   (London, 1873)
Memalphidea or venomous snakes of India   (London, 1874)
On preservation of health in India   (London, 1880)
On the climate and fevers of India   (London, 1882)
Recollections of my life (Edinburgh and London, 1900)

<u>Biographical Studies</u>

Munk, W., Roll of the Royal College of Physicians of London, <u>4</u>, (London, 1878), pp.201-203
Who's Who, 1898, (London 1898)
World Who's Who in Science, (Chicago, 1968), p.550
Comrie, J.D., History of Scottish Medicine, <u>2</u>, (London, 1932), p.761

<u>Manuscript Locations</u>

E; EU; LA

Andrew FLEMING (1822-1901)

Surgeon and Naturalist

Elected F.R.S.E.   6 December 1852

<u>Portraits, etc.</u>

None known

Born on 7 March 1822, little is known of Andrew's early life. After a period with the Indian Medical Service, he studied medicine at Edinburgh University and graduated M.D. in 1843 with the thesis 'On arsenious acid and its poisonous effects'. He then went to India, becoming an assistant surgeon in 1845 and saw active service during the Sikh Wars. He became surgeon in 1859 and in 1865 he was promoted to Surgeon-Major. While in India he also pursued his interest in geology and in time built up a large collection of fossils from the Punjab. He retired from the army after twenty-eight years service, on 27 January 1873 and returned to Edinburgh where he took up residence at 8, Napier Road, Merchiston. He died on 25 March 1901.

<u>Honours and Distinctions</u>

n.d.   Fellow of the Royal College of Surgeons of Edinburgh
1878   Member of the Edinburgh Harveian Society

<u>Publications</u>

Diary of a trip to Pind Dadud Khan and the Salt Range   (London, 1849)
Blood-stains   (n.p., 1859)
Report on geology and mineral wealth of the Salt Range in the Punjab   (n.p., 1853)

<u>Biographical Studies</u>

Crawford, D.G., Roll of the Indian Medical Service   (London, 1930), p.129
Wemyss, H.L.W., A record of the Edinburgh Harveian Society   (Edinburgh, 1833), p.64
List of graduates in medicine in the University of Edinburgh, from 1705 to 1866, (Edinburgh, 1867), p.131

<u>Manuscript Locations</u>

EU; SaU

<u>Collections</u>

EF

John Gibson FLEMING (1809-1879)

Physician and Surgeon

Elected F.R.S.E. 15 January 1872

<u>Portraits, etc.</u>

Portrait: (Sir D. MacNee), Faculty of Physicians
    and Surgeons, Glasgow
Engraving: see Anderson, J.W., Four chiefs of
    Glasgow Royal Infirmary, (Glasgow, 1916) **

Born in Glasgow on 2 December 1809, John
received his early education at the High
School in Glasgow. He then entered Glasgow
University where he studied medicine and at
the age of twenty-one graduated M.D. After a
period of study abroad he returned to Glasgow
and soon built up a large practice. A well-
known city figure, he served for many years as
the representative for the faculty of
Physicians and Surgeons in Glasgow at the General Council of Medical Education and Registration.
His medical writings were few but his "Medical Statistics of Life Assurance" was deemed a
valuable contribution to the medical departments of life assurance firms. He was manager of
several charitable institutions in Glasgow and during his time as surgeon and physician to the
Royal Infirmary, made many improvements. He died of typhoid at 155 Bath Street, Glasgow on 2
October 1879.

<u>Honours and Distincions</u>

1833   Fellow of the Faculty of Physicians and Surgeons of Glasgow
1865-71   President of the Faculty of Physicians and Surgeons of Glasgow

<u>Publications</u>

The pathology and treatment of ramollissement of the brain   (Glasgow, 1833)
Medical statistics of life assurance being an inquiry into the causes of death among the
members of the Scottish Amicable Life Assurance Society, from 1826 till 1860, and a complete
analysis of the diseases which have been proved fatal among the assured in several Societies
(Glasgow, 1862)
Address delivered at the inauguration of the Glasgow Royal Infirmary School of Medicine
(Glasgow, 1876)

<u>Biographical Studies</u>

<u>Proc. R. Soc. Edinb.,</u> <u>10,</u> (Edinburgh, 1880), pp.346-8
Duncan, A., Memorials of the Faculty of Physicians and Surgeons of Glasgow, (Glasgow, 1896),
p.286
Biographisches Lexikon der Hervorragenden Aerzte, <u>2,</u> (Berlin, 1929), p.540
Boase, F., Modern English Biography, <u>1,</u> (London, 1965), p.1068

<u>Manuscript Locations</u>

None known

Sir Thomas Richard FRASER (1841-1920)

Pharmacologist

Elected F.R.S.E.   4 February 1867

Portraits, etc.

Portrait: (R. Home, 1919), EU
Portrait: (T.M. Ronaldson), RCPE
Photograph: see Comrie, J.D., History of Scottish
    Medicine, (London, 1932) **

Born in Calcutta on 5 February 1841, Sir Thomas
received his early education at public schools
in Scotland. After graduating M.D. in 1862 with
a thesis on the Calabar bean, he continued his
research under Sir Robert Christison and in
1869 was appointed Physician to the Royal
Infirmary. Eight years later he succeeded to
the Chair of Materia Medica, a post which he
held for forty years. His early research work
was on the action of arrow and ordeal poisons, and later he dealt with the subject of
antagonisms between different active substances. This led to important work on snake venoms and
anti-venin. For many years he was also involved in examining the remedies obtained from plants
of the strophanthus group and this resulted in the introduction of strophanthin into medicine.
Along with Alexander Crum Brown, he was the first to study the relationship between the
chemical structure of drugs and their physiological action. He was for many years Dean of the
Faculty of Medicine and a member of the University Court and during his lifetime received many
honours from other universities, including Aberdeen, Dublin and Cambridge. In 1898 he was
appointed President of the Indian Plague Commission on which he served for two years. He was
married to Susanna Margaret Duncan by whom he had four sons and three daughters. He retired in
1918 and died in Edinburgh on 4 January 1920.

Honours and Distinctions *

1877   Fellow of the Royal Society of London
1900   President of the Royal College of Physicians of Edinburgh
1901   Honorary LL.D. Glasgow University
1902   Knighted
1904   Honorary Physician to the King in Scotland
1907   Honorary Sc.D. Cambridge University
1908   President of the Association of Physicians of Great Britain and Ireland

Publications *

On the physiological action of the Calabar bean   (Edinburgh, 1867)
"An experimental research on the antagonism between the actions of physostigma and atropia",
Trans. R. Soc. Edinb., 26, (Edinburgh, 1872), pp.529-713
On the action and uses of digitalis and its substitutes with special reference to strophanthus
(London, 1886)

Biographical Studies *

Who's Who, 1898, (London, 1898)
World's Who's Who in Science, (Chicago, [1968]), p.601
Logan-Turner, A., (ed). History of the University of Edinburgh 1883-1933, (Edinburgh, 1933),
pp.113, 382
Comrie J.D., History of Scottish Medicine, 2, (London, 1932), pp.712-3

Manuscript Locations

EPH; EU; SaU

Arthur GAMGEE (1841-1909)

Physician and Physiologist

Elected F.R.S.E.   21 January 1867

<u>Portraits, etc.</u>

Photograph: see Comrie,J.D., History of Scottish
    Medicine, (London, 1932) **

Born in Florence on 10 October 1841, the youngest son of Joseph Gamgee, a veterinary surgeon, and Mary West, Arthur received his early education at the University College School, London. He then studied medicine at Edinburgh University and graduated M.D. in 1862. After this, he became assistant to the Professor of Medical Jurisprudence and also gave lectures on physiology at the Royal College of Surgeons of Edinburgh. Six years later he went to Germany, where he studied under Kuhne at Heidelberg and Ludwig at Leibzig, before returning to Edinburgh as a lecturer in physiology at the extra-mural school. In 1873, he was appointed first Brackenbury Professor of Physiology at Owen's College, Manchester and remained in this post for twelve years. During this time he was also physician to the Hospital for Consumption and from 1882, was Fullerian Professor of Physiology at the Royal Institution. After a period of practice in St. Leonard's, Sussex, he settled in London in 1887 where he was assistant physician and lecturer on pharmacology at St. George's Hospital. However ill-health forced him to resign and so he moved to Switzerland where he set up a consulting practice. He delivered his Croonian lectures in 1902 and subsequently visited America to inspect physiological laboratories there. Besides his medical expertise he was also an accomplished linguist and classical scholar. In 1875 he married Mary Louisa, daughter of J. Proctor Clark and had three children by her. He was the uncle of Sir D'Arcy Wentworth Thomas and brother of Joseph Gamgee (1828-86). He died while on a visit to Paris on 2 March 1909.

<u>Honours and Distinctions</u> *

1872   Fellow of the Royal College of Physicians of Edinburgh
1873   Fellow of the Royal Society of London
1896   Fellow of the Royal College of Physicians of London
1908   Honorary D.Sc., Manchester University

<u>Publications</u> *

On force and matter in relation to organization; an introductory lecture to a course on physiology (Edinburgh, 1869)
Studies from the physiological laboratory at Owen's College (Cambridge, 1877)
The digestive forments and the chemical processes of digestion (London, 1884)
On the behaviour of oxy-haemoglobin, carbonic-oxide-haemoglobin, methaemoglobin and certain of their derivatives in the magnetic field, with a preliminary note on the electrolysis of the haemoglobin compounds (London, 1901)

<u>Biographical Studies</u> *

Munk, W., Roll of the Royal College of Physicians of London, 4, (London, 1878), pp.387-8
Who's Who 1898, (London, 1898)
Comrie,J.D., History of Scottish Medicine, 2, (London, 1932), p.713
Royal College of Physicians, Library Catalogue, 1, (Edinburgh 1898), p533

<u>Manuscript Locations</u>

SaU

Joseph Sampson GAMGEE (1828-1886)

Surgeon, Queen's Hospital, Birmingham

Elected F.R.S.E. 2 March 1868

<u>Portraits, etc.</u>

Photograph: see Hamilton Baily and W.J. Bishop,
    Notable names in medicine and surgery
    (London, 1959) **

Born at Livorno, Italy, on 17 April 1828, the eldest son of Joseph Gamgee, a veterinary surgeon, and Mary West, Joseph received his early education in Florence. After a period of study in Paris, he studied for and obtained his diploma at the Royal Veterinary College London, then entered University College, London where he won the Liston Prize in 1853 for an essay "On the advantages of the starched apparatus in the treatment of fractures". After obtaining his membership of the Royal College of Surgeons of London in 1854, he acted as house-surgeon at the University College Hospital before volunteering his services in the Crimean War as surgeon to the British-Italian Legion in Malta. On return he became assistant surgeon at the Royal Free Hospital and in 1857 was elected one of the honorary surgeons at the Queen's Hospital in Birmingham. He remained here until 1881 and during that time rendered great service to the hospital and the medical school associated with it. He was mainly responsible for starting up the 'Hospital Saturday' collections in Birmingham and during the Franco-Prussian war of 1870-71 turned his surgery into an ambulance depot. His practical skill as a surgeon was well-known and he also introduced many hygienic improvements such as the use of absorbent cotton-wool pads, gauze tissue, millboard and paper splints. He contributed regularly to 'The Lancet' and his linguistic skill allowed him to have good communication with medics abroad and their writings. He retired in 1881 on account of ill-health and died five years later on 18 September 1886. He was married to Marion Parker by whom he had seven children.

<u>Honours and Distinctions</u>

1870-1   Secretary of the Birmingham Society for the Aid to the wounded

<u>Publications</u> *

Account of a calcified testicle of a ram   (London, 1850)
On pyaemia (London, 1853)
Researches in pathological anatomy and clinical surgery (London, 1856)
The cattle plague and diseased meat in their relations with the public health and with the interests of agriculture (London, 1857)
History of a successful case of amputation at the hip joint (London, 1865)
Medical report: the present crisis (London, 1870)
On the treatment of fractures of the limbs (London, 1871)
On the treatment of wounds (London, 1878)
On absorbent and anti-septic surgical dressings (London, 1880)
The influence of vivisection on human surgery (London, 1882)

<u>Biographical Studies</u> *

DNB, (London, 1968)
Edinb. Med. J. 32, (Edinburgh, 1887), pp.479-80
Hamilton Baily and Bishop, W.J., Notable Names in Medicine and Surgery, (London, 1959), pp.105-6

<u>Manuscript Locations</u>

SaU

Robert Mortimer GLOVER (1816-1859)

Physician

Elected F.R.S.E.   2 December 1850

<u>Portraits, etc.</u>

None known

Born in London in 1816 little is known of Glover's early life. He appears to have moved to Edinburgh at an early age. He studied medicine at Edinburgh University and graduated M.D. in 1840 with the thesis 'On the physiological and medicinal properties of bromine and its compounds'. He then set up practice in Newcastle-upon-Tyne and in 1842 was awarded a medal by the Harveian Society for his essay 'Experimental enquiry into the physiological and therapeutical properties of bromine and its compounds'. He died on 9 April 1859 at 1 Kensington Park Road, London from an overdose of chloroform.

<u>Honours and Distinctions</u>

1842   Harveian Prize

<u>Publications</u> *

Substance of a lecture on the applications of chemistry to medicine (Gateshead, 1842)
On the physiological and medicinal properties of bromine and its compounds (Edinburgh, 1842)
On the pathology and treatment of Scrofula being the Fothergillian Prize for 1846 (London, 1846)
On the philosophy of medicine on quackery (London, 1851)
On the physiological properties of Picrotoxin (Edinburgh, 1851)
A manual of elementary chemistry (London, 1855)
The mineral waters: their physical and medicinal properties (London, 1857)
The New Medical Act (London, 1858)
Acute pneumonia, not a fatal disease; its therapeutics (London, 1862)

<u>Biographical Studies</u> *

Wemyss, H.L., Harveian Society Record, 1782-1933, (Edinburgh, 1933), p.24
Biographisches Lexikon der Hervorragenden Aerzte, <u>2</u>, (Berlin 1929), p.772
Comrie, J.D., History Scottish Medicine, <u>2</u>, (London 1932), p.628

<u>Manuscript Locations</u>

SaU

Andrew GRAHAM   (1817-1889)

Fleet Surgeon, Royal Navy

Elected F.R.S.E.   21 January 1867

<u>Portraits, etc.</u>

None known

Born in Dalkeith in 1817, the son of Walter Graham M.D., who died in 1827, Andrew received his early education locally. He then went to Edinburgh University to study medicine and graduated M.D. in 1838 with the thesis 'On Aneurism'. He then entered the navy as an assistant surgeon and on 27 July 1847 became a staff surgeon. He saw active service in the Mediterranean and East Indies and during the Russian War served on board a flagship in the Baltic under Admiral Sir Charles Napier. At a later date, he served in the West Indies aboard 'H.M.S. Agamemnon' under Captain Thomas Hope Pinkie and for his services, he was awarded the Baltic Medal and Sir Gilbert Blane's gold medal. He was appointed fleet surgeon on 22 June 1864 and remained in the post for six years. He died at his home at the Albany, Piccadilly, London on 1 December 1889.

<u>Honours and Distinctions</u>

1869   Fellow of the Royal Geographical Society

<u>Publications</u>

Medical topography of Singapore and Sarawak   (Edinburgh, 1852)

<u>Biographical Studies</u>

<u>Proc. R. Soc. Edinb., 17,</u> (Edinburgh, 1890), p.25
List of the graduates in medicine in the University of Edinburgh from 1705 to 1866, (Edinburgh, 1867), p.114
The Medical Directory for 1890, (London, 1890), p.1632

<u>Manuscript Locations</u>

None known

Daniel Rutherford HALDANE (1824-1887)

Physician

Elected F.R.S.E   4 February 1867

<u>Portraits,etc.</u>

Photograph: Medical Archives Centre, Royal
    Infirmary of Edinburgh **

Born in Edinburgh in 1824, the son of James
Alexander Haldane and Margaret Rutherford,
Daniel received his early education at the
Edinburgh High School, and then studied
medicine at the University. He graduated M.D in
1848 and was awarded a gold medal for his
thesis on 'Diseases of the Liver'. He became a
resident physician at the Royal Infirmary for a
while, before travelling to Vienna and Paris
for further study. When he returned to
Edinburgh in 1854 he secured the same post and
was connected with the clinical department of Edinburgh University. Shortly after he became
physician to the Royal Public and New Town Dispensaries, as well as lecturer on Medical
Jurisprudence, teacher of pathology and morbid anatomy in the extra-mural school at Surgeon's
Hall, and pathologist to the Royal Infirmary. Haldane was an active member of the Royal College
of Surgeons, being secretary for sixteen years and then president. It was Haldane who arranged
for a common diploma examination to be established by the Royal College of Surgeons and
Physicians of Edinburgh, and the Faculty of Physicians and Surgeons of Glasgow. Every medical
student had to sit this examination. For this he was awarded a silver plate. He was a
representative of the Royal College of Physicians at the General Council of Medical Education,
assessor of the General Council of Edinburgh University at the University Court, and in 1875 was
appointed examiner in clinical medicine by that body. In later years he was consulting physician
to the Royal Infirmary and editor of the 'Edinburgh Medical Journal'. He died in Edinburgh on
12 April 1887.

<u>Honours and Distinctions</u>

1846   Senior President of the Royal Medical Society, Edinburgh
1852   Fellow of the Royal College of Physicians of Edinburgh
1879   President of the Royal College of Physicians of Edinburgh
1880   President of the Harveian Society, Edinburgh
1883   LL.D., Edinburgh University

<u>Publications</u>

Introductory address, delivered at the Medical School, Surgeon's Hall, Edinburgh (Edinburgh, 1857)
Cellular pathology; case of syphilitic deposit in the substance of the heart (Edinburgh, 1862)
The modern practice of medicine (Edinburgh, 1865)

<u>Biographical Studies</u>

DNB, (London, 1968)
<u>Edinb. Med. J.</u> 32, (2), (Edinburgh, 1887), pp.1060-3
Gray, J., History of the Royal Medical Society 1737-1937 (Edinburgh, 1952), p.178
Biographisches Lexikon der Hervorragenden Aerzte, 3, (Berlin, 1929), p.25
Royal College of Physicians of London, Library Catalogue, (London, 1912), p.540

<u>Manuscript Locations</u>

E; EU; SaU

James Herbert Brockencote HALLEN   (1829-1901)

Inspecting Veterinary Surgeon to H.M. Indian
Army

Elected F.R.S.E.   18 March 1867

<u>Portraits, etc.</u>

None known

Born in 1829, the son of an army veterinary
surgeon, John received his early education at
Bridgenorth and Leeds Grammar School. He then
went to Edinburgh to study veterinary science
under Professor Dick and medicine at the
University, where he graduated M.D. In 1850 he
went to India where he spent the rest of his
professional life in the veterinary services of
the British Army. He served for five years with
the 1st Bombay Lancers in Rajputana, was
employed on remount purchasing duties in Bombay
for a year from 1857, and then served with the Bombay Horse Artillery for four years. During
that time he established and ran a regimental veterinary school at the headquarters of the
Bombay Artillery. In 1862 he was made principal veterinary surgeon of the Bombay Army, and a
year later was nominated Superintendent of the Horse Breeding Department of the Bombay
Presidency. Throughout the Abyssinian campaign (1867-8) he served as principal veterinary
surgeon of the force, being in charge of ten vets and over five thousand animals of the cavalry,
artillery and transport corps. In 1868 he was appointed officiating Inspecting Veterinary
Surgeon of the Bombay Army, and the following year he became President of the Indian Cattle
Plague Commisssion, continuing in this capacity until 1872. He was then appointed Acting
Inspecting Veterinary Surgeon, and was asked by the Governors of India to organize a Veterinary
College in Calcutta, being made Principal of the proposed College. While engaged in those duties
his services were requested by the Viceroy in Council on the Special Stud Commmission, and he
continued with the Commission until 1876, when he became Government Superintendent of Horse
Breeding. In this post he did much to improve local horse breeds, and instituted a veterinary
school in the Punjab. Hallen retired in 1894, returning to England sometime after that. He died
at his home at Pebworth Fields, near Stratford-on-Avon on 1 August 1901.

<u>Honours</u> <u>and</u> <u>Distinctions</u>

None known

<u>Publications</u>

Manual of the more deadly forms of cattle diseases in India, 1871   (n.p. 1872)

<u>Biographical</u> <u>Studies</u>

<u>Proc. R. Soc. Edinb.,</u> <u>24,</u>   (Edinburgh, 1903), pp.642-5
Who's Who, 1900, (London, 1900)
Who Was Who, <u>1,</u> 1897-1915, (London, 1967)
NUC, <u>227,</u> p.578
The Medical Directory for 1902, (London, 1902), p.2116

<u>Manuscript</u> <u>Locations</u>

None known

William Bird HERAPATH (1820-1868)

Surgeon and Toxicologist

Elected F.R.S.E.   17 April 1854

<u>Portraits, etc.</u>

None known

William Herapath was born in Bristol on 28 February 1820, the eldest son of William Herapath, an analytical chemist and Professor of Chemistry and Toxicology at the Bristol Medical School. William studied medicine at London University and in 1843 he was made a Licenciate of the Society of Apothecaries, and M.B. the following year. He was elected to the Royal College of Surgeons and began work at Queen Elizabeth's Hospital in Bristol. In 1851 he received his M.D.. Herapath published many articles in scientific journals, in which he reported a number of important discoveries. One was the production of small but usable crystals of iodosulphate of quinine, now known as 'herapathite'. He also devised new methods for detecting arsenic and other substances and designed a new combustion blowpipe for organic analyses. He used the spectroscope and microspectroscope to detect bloodstains, developed new techniques for pathological investigations and experimented with alkaloids. Herapath was also interested in subjects of a less purely scientific nature, such as domestic sanitation and the analyses of spa waters and hot wells. By 1864 he had fallen ill with a form of jaundice, but despite this, continued his work on spectroscopic analysis until a few days before his death on 12 October 1868.

<u>Honours</u> <u>and</u> <u>Distinctions</u>

1844   Fellow of the Royal College of Surgeons of England
1859   Fellow of the Royal Society of London

<u>Publications</u>

On the chemical constitution and atomic weight of the new polarizing crystals produced from quinine (London, 1852)
A few words on the Bristol and Clifton Hot Wells. Together with an analysis of the spa (Bristol, 1854)
The importance of the presentation of breech or inferior extremity in cases of infanticide (London, 1859)
Address on chemistry on its relations to medicine and its collateral sciences (Bristol, 1863)

<u>Biographical</u> <u>Studies</u>

DSB, (New York, 1976)
<u>Illus.</u> <u>Lond.</u> <u>News,</u> 24 October 1868, (London, 1868)
Royal College of Physicians of London, Library Catalogue, (London 1912), p.581
BMC, <u>102,</u> 1961, p.136
NUC, <u>241,</u> p.451

<u>Manuscript Locations</u>

SaU

Charles Hayes HIGGINS (1811-1898)

Surgeon

Elected F.R.S.E. 30 January 1871

<u>Portraits, etc.</u>

Photograph: see History of the Birkenhead
    Library and Scientific Society, 1857-1907,
    (privately published by the Society, 1907) **

Born in 1811 on board the flagship of Admiral
Sir John Hayes in the habour of Batavia,
Charles remained in India until the age of nine
and then returned to England, attending a
school in Hammersmith and Dr Radcliffe's school
at Salisbury. He studied medicine at Guy's
Hospital, London, becoming a member of the
Royal College of Surgeons and Licentiate of the
Society of Apothecaries in 1834. Subsequently
he spent some months in Paris, then at the
University of Edinburgh. He settled at Taunton and became surgeon to the Somerset County
Hospital. An accident to his right hand virtually ended his career as a surgeon and in 1850 he
moved to Birkenhead, remaining there for the rest of his life. He was Surgeon to the Borough
Hospital there, Senior Consulting Physician to the Wirral Children's Hospital and Physician to
the Birkenhead Eye and Ear Hospital and Dispensary. In addition to this, he was Surgeon-Major in
the 1st Cheshire Engineer Volunteers. Higgins was also active in local literary and scientific
life, being twice President of the Birkenhead Medical Association, and also twice President of
the Birkenhead Literary and Scientific Society. He was much respected in Birkenhead and was the
senior medical man in his later years. He died in Birkenhead on 14 January 1898.

<u>Honours</u> <u>and</u> <u>Distinctions</u>

1844   Fellow of the Royal College of Surgeons of England
1849   Fellow of the Obstetric Society of Edinburgh

<u>Publications</u>

Observations on climate, diet and medical treatment in France and England (London, 1835)
Notes sur l'emploi des alterants dans les maladies aigues et chroniques (Paris, 1859)
'On compression in certain surgical affections', <u>Liverpool</u> <u>Medical</u> <u>Transactions,</u> <u>2</u>
'On diagnosis of fractured clavicle', <u>Medical</u> <u>Gazette</u>
'Case of inversion of uterus from polypus - extirpation successfully performed with knife',
<u>Edinb.</u> <u>Med.</u> <u>Surg.</u> <u>J.</u> (Edinburgh, 1849)
'Nature and use of uterine sound', <u>Medical</u> <u>Times</u> <u>Gazette,</u> 1854

(The above information is taken from the Medical Directory for 1898, p.813)

<u>Biographical</u> <u>Studies</u>

Plarr, V.G., Lives of the Fellows of the Royal Society of England, <u>1,</u>
(Bristol and London, 1930), p.535
BMC, <u>103,</u> 1961, p.700
The Medical Directory for 1899, (London, 1899), p.1952

<u>Manuscript</u> <u>Locations</u>

EU

Alexander HUNTER (1816-1890)

Surgeon-Major, Madras Army

Elected F.R.S.E.   2 March 1874

<u>Portraits, etc.</u>

None known

Born on 19 May 1816, little is known of Hunter's early life. He studied medicine at Edinburgh University, graduating M.D. in 1837 with the thesis 'On the pathology and treatment of granular disease of the kidney'. He joined the Indian Medical Service as Assistant Surgeon in February 1843, rising to the rank of Surgeon-Major in February 1863. He retired from the Service in 1873. Hunter died in 1890.

<u>Honours and Distinctions</u>

None known

<u>Publications</u>

A probationary essay on phlebitas, submitted .... to the examination of the Royal College of Surgeons of Edinburgh, when candidate for admission into their body. (Edinburgh, 1839)

<u>Biographical Studies</u>

BMC, <u>109</u>, 1961, p.448
List of the graduates in medicine in the University of Edinburgh, from 1705 to 1866, (Edinburgh, 1867), p.111
Comrie, J.D., History of Scottish Medicine, <u>2</u>, (London, 1932), p.629
<u>Proc. R. Soc. Edinb.</u>, <u>9</u>, (Edinburgh, 1878), p.488
Crawford, D.G., Roll of the Indian Medical Service, (London, 1930), p.336

<u>Manuscript Locations</u>

SaU

Alexander KEILLER   (1811-1892)

Gynaecologist

Elected F.R.S.E.   4 December 1865

<u>Portraits, etc.</u>

Portrait: (A.A. Gambley, 1875-77), RCPE
Photograph: see Comrie, J.D., History of Scottish
    Medicine, (London, 1932) **

Born in Arbroath on 11 November 1811, Alexander
received his early education from Mr. David
Lindsay of Dundee. He studied medicine in
Edinburgh and St. Andrew's University,
graduating M.D. from the latter in 1835. He was
one of Robert Knox's favourite pupils, and when
the latter gave up lecturing in 1842, Keiller
studied anatomy with Dr. Skae. After two
sessions he moved to Dundee and stayed there
for seven years as a general practitioner. In
1851 he went back to Edinburgh to take up an appointment as ordinary physician to the Royal
Infirmary, which he held for fifteen years. He lectured on materia medica, medical
jurisprudence, and on midwifery and the diseases of women and children. The last category became
his favourite subject, and the one to which he devoted the rest of his life. He arranged a
course on clinical teaching on the diseases of women, and an extra ward was set apart solely for
this purpose. This was the beginning of gynaecological teaching in the Edinburgh Medical School.
Keiller was also physician to the Edinburgh Maternity Hospital, and was examiner in midwifery
for St. Andrew's University, Edinburgh University and the Edinburgh Colleges of Physicians and
Surgeons. He also lectured on the diseases of children at the Royal Infirmary of Edinburgh. He
invented various obstetrical instruments, and was honorary fellow of the Obstetrical Society of
London and of the Gynaecologist Society of Boston, U.S.A. In 1846, he married Miss Roy, daughter
of Major Roy of the Bengal service and had three children by her. He died at North Berwick on 26
September 1892.

<u>Honours and Distinctions</u> *

1849   Fellow of the Royal College of Physicians of Edinburgh
1860   President of the Obstetrics Society
n.d.   Honorary Fellow of the Obstetrical Society of London
1875   President of the Royal College of Physicians of Edinburgh
1886   Honorary LL.D., St. Andrew's University

<u>Publications</u>

On spurious pregnancy and hysteria   (Edinburgh, 1855)
Medical-legal observations on manual strangulation   (Edinburgh, 1855)
Introductory address, delivered at the opening of session 1858-9 at the Medical and Surgical
School, Surgeons' Hall (Edinburgh, 1859)

<u>Biographical Studies</u> *

<u>Edinb. Med. J.</u>, <u>38</u>, (1), (Edinburgh, 1876), pp.491-4
<u>Proc. R. Soc. Edinb.</u>, <u>19</u>, (Edinburgh, 1893), pp.xxxiv-xxxix
Comrie, J.D., History of Scottish Medicine, <u>2</u>, (London, 1932), pp.604-5, 616, 626, 628-9, 631,
685
Boase, F., Modern English Biography, <u>5</u>, (suppl.), (London, 1965), p.803
<u>Br. med. J.</u>, <u>2</u>, 1895, (London, 1895), p.795

<u>Manuscript Locations</u>

EU; SaU

Thomas LAYCOCK   (1812-1876)

Mental Physiologist, Professor of the Practice
of Physic at the University of Edinburgh

Elected F.R.S.E.   18 February 1856

<u>Portraits, etc.</u>

Portrait: (M. Hardie, 1856), RCPE
Photograph: see Comrie, J.D., History of Scottish
     Medicine, (London, 1932) **
Bust: (C. Stanton, RSA), EU
Photograph: see MSS. Cat., EUL

Born at Wetherby, Yorkshire on 10 August 1812,
son of the Rev. Thomas Laycock, a Wesleyan
minister, Thomas was educated at the Wesleyan
Academy, Woodhouse Grove and University
College, London, where he studied medicine. He
continued his studies in Paris and Gottingen,
graduating M.D. from the latter in 1839. He
then went to York and settled there as a medical practitioner, where after a few years he was
appointed to the Chair of Medicine at the York Medical School. He held this post until 1855,
when he moved to Edinburgh to take up the Chair of the Practice of Physic at the University,
which he held until his death. In 1869 he was appointed Physician-in-Ordinary to the Queen in
Scotland, being the only Englishman ever to have been elected to this position, also retaining
this until his death. Early on in his career, Laycock began studying the relations between the
nervous systems and psychological phenomena. He was the first to formulate the theory of the
reflex action of the brain (1844) by which he accounted for the phenomenon of delirium, of
dreams and somnabulism, and also first to apply the theory of evolution to the development of
the nervous centres in people and animals. Laycock was a prolific writer on medical subjects,
contributing between two and three hundred papers to medical and scientific journals and was
editor and translator of various works, such as J.A. Unger's "Principles of Physiology". All his
professional life he was absorbed in research and teaching, although he was not considered
popular in this last capacity. He died at his home in Walker Street, Edinburgh on 21 September
1876.

<u>Honours and Distinctions</u>

1835   Member of the Royal College of Surgeons of England
1844   Secretary to the British Association
1856   Fellow of the Royal College of Physicians of Edinburgh
1861   Fellow of the Royal Society of London
1869   Physician-in-Ordinary to the Queen in Scotland

<u>Publications</u> *

A treatise on the nervous diseases of women; comprising an inquiry into the nature, causes, and
treatment of spinal and hysterical disorders (London, 1840)
On the sanitary conditions of the city of York   (York, 1844)
Report on the state of York   (York, 1844)
Lectures on the principles and methods of medical observation and research (Edinburgh, 1856)
The social and political relations of drunkenness: two lectures (Edinburgh, 1857)
Mind and brain, or, The correlations of conciousness and organization, with their applications
to philosophy, zoology, physiology, mental pathology, and the practice of medicine
(Edinburgh, 1860)
An etiological nosology of diseases of the skin (Edinburgh, 1862)
The principles and methods of medical observation and research. 2nd. ed., with copious
nosologies and indexes of fevers (Edinburgh, 1864)

<u>Biographical Studies</u> *

BMC, <u>132</u>, 1962, p.134
DNB, (London, 1968)
Amacher, M.Peter, "Thomas Laycock, J.M. Sechenov and the reflex arc concept",
<u>Bull. Hist. Med.</u>, <u>38</u>, (Baltimore, 1964), pp.68-83
<u>Edinb. Med. J.</u> <u>22</u>, (Edinburgh, 1876), pp.476-9
<u>Proc. R. Soc. Edinb.</u>, <u>9</u>, (Edinburgh, 1878), pp.223-5
Comrie, J.D., History of Scottish Medicine, <u>2</u>, (London, 1932), pp.612-613

<u>Manuscript Locations</u>

E; EPH; EU; L; LUC; SaU

William Lauder LINDSAY (1829-1880)

Alienist and Botanist

Elected F.R.S.E.   4 February 1861

<u>Portraits, etc.</u>

None known

Born in Edinburgh on 19 December 1829, the eldest son of James Lindsay of H.M. Sasine Office, Register House, William received his early education at the local high school, where he was dux in 1844-5. He then studied medicine at the University and graduated M.D. with the highest honours in 1852 with his thesis 'Anatomy, morphology and physiology of the Lichens'. A year later he went to Dumfries to become assistant physician to the Crichton Asylum, and the following year moved to Perth as physician to the Royal Murray Asylum, where he remained for twenty-five years. While in Perth, Lindsay made lichens his special study, and he also pursued his botanical researches abroad, travelling to New Zealand in 1861-2, as well as visiting North Germany, Iceland, the Faroes and Norway. The results of these studies were published in his many botanical works and papers. He also wrote on medicine, natural history and received the Neill gold medal from the Royal Society of Edinburgh for original investigations in the structure of lichens. In November 1879 he had to retire from the Royal Murray Asylum due to ill-health, and died a year later, at 3 Hartington Gardens, Edinburgh on 24 November 1880, leaving a widow and a daughter.

<u>Honours and Distinctions</u>

1858   Fellow of the Linnean Society
1859   First Neill gold medallist of the Royal Society of Edinburgh

<u>Publications</u> *

The histology of the blood in the insane (Perth, 1854)
A popular history of British Lichens (London, 1856)
The flora of Iceland (Edinburgh, 1861)
Contributions to New Zealand Botany (London and Edinburgh, 1868)
The physiology and pathology of mind in the lower animals (Edinburgh, 1871)
The development and reform of the higher education in Scotland (Perth, 1874)
The theory and practice of non-restraint in the treatment of the insane (Edinburgh, 1878)
General history of the Royal Murray Institution in Perth (Perth, 1878)

<u>Biographical Studies</u>

DNB, (London, 1968)
<u>Edinb. Med. J.</u>, <u>26</u>, (2), (Edinburgh, 1881), pp.669-72
<u>Trans. Bot. Soc. Edinb.</u>, <u>14</u>, (Edinburgh, 1883), pp.163-4
Britten, J. & Boulger, G.S., A biographical index of deceased British and Irish Botanists, 2nd ed., (London, 1931), p.387
Boase, Frederic, Modern English Biography, <u>2</u>, (London, 1965), p.91

<u>Manuscript Locations</u>

EU; L;

Lord Joseph LISTER (1827-1912)

Professor of Clinical Surgery, King's College, London

Elected F.R.S.E. 3 January 1870

<u>Portraits, etc.</u>

Portrait: (J.H. Lorimer, R.S.A., 1895), EU
Photograph: see Comrie, J.D., History of Scottish
    Medicine, (London, 1932)
Bust: Westminster Abbey, London
Photograph: Medical Archives Centre, Royal
    Infirmary of Edinburgh **

Joseph Lister was born at Upton House, Plaistow on 5 April 1827, the second son of Joseph Jackson Lister, F.R.S., a London wine-merchant and inventor of the achromatic microscope. He was educated at private schools in Hitchin and Tottenham before entering University College, London in 1844 to study Arts. He graduated B.A. three years later, and then went on to study medicine, graduating in 1852. The following year he came to Edinburgh where he worked as a house-surgeon under James Syme. In 1856 he married Syme's eldest daughter, Agnes and in that same year he became assistant surgeon to the Royal Infirmary, but gave this up to take the Chair of Surgery at the University of Glasgow in 1860, returning to Edinburgh nine years later as Professor of Surgery. While in Glasgow, Lister invented several useful surgical instruments including a sinus forceps, an abdominal tourniquet, probe-pointed scissors and an ear-hook. He also devised a new technique for incisions of the wrist joint in cases of tuberculous disease. Earlier he had published research results on the coagulation of the blood and inflammation, which were important in relation to the problem of wound infection but it is as the father of antiseptic surgery that his name is famous throughout the world. While the discovery of general anaesthesia in 1846 had taken the pain out of surgery, mortality had increased; Lister helped make surgery safer. Through the work of Pasteur, his attention was drawn to germ theory. He became convinced that the air was the main source of infection, and devised carbolic sprays to keep it at bay. He also invented absorbent gauze dressing, experimenting for some time to find the ideal formula. This he discovered in 1865, and published his results in 'The Lancet' a year later. Using carbolic acid as a germicide, he lost only fifteen per cent of his patients after amputation; before, the figure had been forty-three per cent. He gained a large following of colleagues and students not only in Scotland, but in America, France and Germany his work was also highly appreciated. At fifty he was offered the Chair of Clinical Surgery at King's College, London, and accepted it with the sole intention of making known his theories in London, where almost no attention had been paid to them. There he met with the indifference and hostility of students and colleagues alike, but did not give up, and during his fifteen years at the College, his reputation steadily increased. In later life he was awarded many honours and was the first surgeon ever to have been made a peer. He represented the Royal Societies of Edinburgh and London at the Sorbonne at the celebration of Pasteur's seventieth birthday. After the death of his wife in 1893, he gradually withdrew from public life. He spent his last years at Walmer, Kent, where he died on 10 February 1912.

<u>Honours and Distinctions</u> *

1860	Fellow of the Royal Society of London
1883	Created Baronet
1896	President of the British Association
1897	Created Peer
1895-1900	President of the Royal Society of London

<u>Publications</u> *

On the coagulation of the blood   (London, 1863)
On the antiseptic principle in the practice of surgery   (London, 1867)
On the effects of the antiseptic system of treatment upon the salubrity of surgical hospital (Edinburgh, 1870)
On the nature of fermentation   (London, 1878)
On the present position of antiseptic surgery   (New York, 1890)
A contribution to the germ theory of putrefaction and other fermentative changes, and to the natural history of torulae and bacteria (Edinburgh, 1873)
"Contributions to physiology and pathology", <u>Phil. Trans. R. Soc.</u>, 1858, (London, 1859)
The collected papers of Baron Joseph Lister, 2 vols., (Oxford, 1909)

<u>Biographical Studies</u> *

Cameron, Hector Charles, Joseph Lister, the friend of man   (London, 1948)
Cartwright, Frederick F., Joseph Lister, the man who made surgery safe (London, 1963)
Godlee, Rickmann John, Joseph Lister, 3rd edition, (Oxford, 1924)
Guthrie, Douglas, Joseph Lister: his life and doctrine   (Edinburgh, 1949)
Thompson, C.J.S., Joseph Lister, the discoverer of antiseptic surgery (London, 1934)
Truax, Rhoda, Joseph Lister, father of modern surgery   (London, 1947)
Walker, Kenneth, Joseph Lister   (London, 1956)
NUC, <u>334</u>, pp.441-3
The London and Provincial Medical Directory, 1859, (London, 1859), p.91
Who's Who, 1898, (London, 1898)
Comrie, J.D., History of Scottish Medicine, <u>2</u>, (London, 1932), pp.635-637, 669

<u>Manuscript Locations</u>

BU; ERM; ES; EU; GC; GF; KM; L; LCP; LIC; LR; LS; LUC; LWM; O

Cosmo Gordon LOGIE (1820–1886)

Surgeon-Major of the Royal Horse Guards

Elected F.R.S.E.   6 February 1871

<u>Portraits, etc.</u>

None known

Little is known of Logie's early life. He was born on 25 August 1820 and studied medicine at Edinburgh University, where he graduated M.D. in 1840 with the thesis "On the total and partial division of the prostate gland in the lateral operation of lithotomy". His career began in the medical service of the British Army as assistant surgeon to the Forces, from 1841 to 1852. He then became surgeon to the Forces, promoted to Surgeon-Major of the Royal Horse Guards in 1861. Logie retired from medical service with half pay and with the honorary rank of Deputy Surgeon General on 30 October 1875. He died in London on 6 April 1886.

<u>Honours and Distinctions</u>

None known

<u>Publications</u>

On the cattle disease   (London, 1866)

<u>Biographical Studies</u>

Johnston, William., Roll of commissioned officers in the medical service of the British Army, 20 June 1727 - 23 June 1898, (Aberdeen, 1917), p.317
BMC., <u>140</u>, (1962), p.956
List of graduates in medicine in the University of Edinburgh, from 1705–1866, (Edinburgh, 1867), p.122

<u>Manuscript Locations</u>

E

William Henry LOWE (1814-1900)

Physician

Elected F.R.S.E.   19 February 1849

<u>Portraits, etc.</u>

None known

Little is known of Lowe's early life. He received his medical education at Edinburgh University, graduating M.D. in 1840. In the same year he was admitted as a member of the Royal College of Physicians of Edinburgh. In 1839 he took membership of the Royal College of Surgeons of England. In 1846 he was elected a Fellow of the Royal College of Physicians of Edinburgh. Dr. Lowe was President of the Royal Medical Society in Edinburgh. He was also Vice-President of the Royal Botanical Society, later becoming President. In 1875 he was President of the Section of Psychology of the British Medical Association. For several years he was resident physician of Saughtonhall Institute for the Insane, and subsequently he ran his own private practice at Balgreen, Murrayfield, Edinburgh. In 1873 he became President of the Royal College of Physicians of Edinburgh. He was one of the doctors who signed a policy for the care of those with incurable diseases. In 1875 Lowe moved to Wimbledon, where he died at the age of eighty-six at his residence, Woodcote, Wimbledon Park on 26 August 1900.

<u>Honours and Distinctions</u> *

1846   Fellow of the Royal College of Physicians of Edinburgh
1873-75   President of the Royal College of Physicians of Edinburgh
n.d.   President of the Botanical Society of London

<u>Publications</u>

Jaundice from non-elimination, with remarks upon the composition and pathology of the bile (Edinburgh, 1840)

<u>Biographical Studies</u>

Desmond, R., Dictionary of British and Irish Botanists and Horticulturalists
(London, 1977), p.396
<u>Br. med. J. 2.</u> (London, 1900), p.700
Craig, W.S., History of the Royal College of Physicians of Edinburgh, (Oxford, 1976), p.494
Royal College of Physicians, Library Catalogue. (London, 1912), p.760

<u>Manuscript Locations</u>

None known

James McBAIN (1807-1879)

Surgeon and Naturalist in the Royal Navy

Elected F.R.S.E. 7 January 1861

<u>Portraits, etc.</u>

None known

Born at Logie, Angus, in November 1807, James received his early education at the Kirriemuir parish school, and after three years as an apprentice to a local surgeon, went to Edinburgh University to continue his medical studies in 1823. Three years later, he passed his medical examination at Surgeons' Hall and received his diploma. He then stayed about a year in St. Andrews, until he was appointed assistant-surgeon to 'H.M.S. Undaunted' in 1827. He sailed with her to India, the Azores and Cape de Verde Islands. Four years later, in the same capacity, he joined a surveying ship, the 'Investigator', assisting in the task of surveying the Shetland and Orkney Islands. The joint efforts of McBain and Captain Thomas in dredging deep waters between the islands produced some interesting results, which were communicated to the naturalists, Forbes, Hanley and Harvey. After sixteen years of survey work, McBain settled in Elie, Fife, and then at Leith and Trinity, Edinburgh, all the time continuing to investigate the marine fauna of the Firth of Forth. He contributed to Rev. W. Wood's topographical work the 'East Neuk of Fife', a catalogue of the Mollusca of the Forth, presenting 344 species, having collected 244 of them himself. He took an active part in the activities of the Royal Physical Society, and contributed many papers from 1859 until his death. He died at Trinity, Edinburgh on 21 March 1879.

<u>Honours and Distinctions</u>

1861-64   President of the Royal Physical Society

<u>Publications</u>

None known

<u>Biographical Studies</u>

<u>Proc. R. Soc. Edinb., 10,</u> (Edinburgh 1880), pp.350-2

<u>Manuscript Locations</u>

BM(IH);   SaU

Thomas Smith MACCALL (c.1804-1895)

Physician

Elected F.R.S.E.   6 April 1868

<u>Portraits, etc.</u>

None known

Little is known of Maccall's early life. He graduated M.D. from St. Andrews University in 1838. He received several honorary titles from Edinburgh bodies, including a Fellowship from the Royal College of Physicians in 1845. His career though seems to have been rooted in Glasgow. He is known to have held the positions of House Physician at Glasgow Royal Infirmary and that of House Surgeon at Lock Hospital, as well as being a junior demonstrator of anatomy at Glasgow University. His most distinguished position was that of Surgeon Major with the 3rd Battalion of the King's Own Royal Lancers Regiment (Glasgow). Maccall lived at Hay-park, Polmont in Stirlingshire after his retirement.

<u>Honours and Distinctions</u>

1831   President of the Hunterian Society of Edinburgh
1845   Fellow of the Royal College of Physicians of Edinburgh
n.d.   Justice of the Peace for Stirlingshire
n.d.   Fellow of the Royal Geological Society

<u>Publications</u>

'Treatment of incontinence of urine', <u>The Lancet</u>, (London, 1872)

<u>Biographical Studies</u>

<u>Proc. R. Soc. Edinb.</u>, <u>10</u>, (Edinburgh, 1880), p.4
Medical Directory for 1885, (London, 1885), p.1056
Royal College of Physicians of Edinburgh, Historical sketch and laws from its institution to 1925, (Edinburgh, 1925), p.10

<u>Manuscript Locations</u>

None known

Angus MACDONALD (1836-1886)

M.D., Obstetrician

Elected F.R.S.E.   30 January 1871

<u>Portraits, etc.</u>

Photograph: see Comrie, J.D., History of Scottish
    Medicine, (London, 1932) **

Born on 18 April 1836 in Aberdeen, the son of
James MacDonald, a road contractor, MacDonald's
early education was mostly self-taught, though
he did attend a parish school for two years. He
then went to King's College in 1855 where he
graduated in 1859, being awarded the Hutton
Prize for general excellence. He had plans to
become a minister and started a course of
theology classes in Edinburgh, but he changed
his ideas and went into medicine in 1860. He
graduated in 1864 and set up his practice in
Edinburgh. He first became a lecturer of materia medica and then midwifery at the Medical
School. He was not primarily interested in the former subject, although he revised Dr. Scoresby-
Jackson's text-book on materia medica in 1875. He later went into obstetrics and gynaecology,
before becoming a physician and clinical lecturer on the diseases of women in the Royal
Infirmary. He was also a physician to the Royal Maternity Hospital. He wrote "The Relation of
Chronic Heart Disease to Pregnancy, Parturition and Childbed", which was the best work of his
published papers. In 1878 he gave lectures on medicine at the extra-mural school in Edinburgh.
MacDonald caught pleuro-pneumonia in 1882, and went to stay on the Riviera. He returned to
Edinburgh not fully recovered, and died on 10 February 1886.

<u>Honours and Distinctions</u>

1875   Fellow of the Royal College of Surgeons
1875   Fellow of the Royal College of Physicians
1880-81   President of the Obstetrical Society

<u>Publications</u> *

The bearing of chronic heart disease upon pregnancy, parturition and childbed (London, 1878)
Revised Dr. Scoresby-Jackson's note book of Materia Medica (Edinburgh, 1871)
"Nature and mechanism of spontaneous rupture of the uterus", <u>Edinb. Med. J.</u>, 1877,
(Edinburgh, 1877)
"Risks and treatment of intra-uterine hydrocephalus", <u>Edinb. Med. J.</u>, 1878, (Edinburgh, 1878)

<u>Biographical Studies</u>

<u>Edinb. Med. J. 31,</u> (Edinburgh, 1886), pp.894, 990-8
Comrie, J.D., History of Scottish Medicine, <u>2,</u> pp.630, 685, 703, 711, 712

<u>Manuscript Locations</u>

E

John Gray McKENDRICK (1841-1926)

Professor of Physiology, Emeritus Professor at the University of Glasgow

Elected F.R.S.E. 7 April 1873

<u>Portraits, etc.</u>

Portrait: (J.H. Lorimer, c.1908), University of Glasgow
Portrait: (Unknown), Townhall, Stonehaven
Photograph: see Comrie,J.D., History of Scottish Medicine, (London, 1932) **

Born at Old Machar, Aberdeen on 12 August 1841, the son of James McKendrick, a merchant, John studied medicine at Aberdeen University graduating M.D. in 1863. Six years later he went to Edinburgh to be assistant to Professor H. Bennett, and in 1872, became a lecturer in the extra-mural school. In 1876 he succeeded Andrew Buchanan as Regius Professor of Physiology at the University of Glasgow. He found the department in poor shape, so immediately began making improvements, including many at his own expense. He organized a well-equipped laboratory and increased practical teaching. He was regarded as a great teacher and was very popular as a lecturer. The consequent increase in the number of students meant that new accommodation was needed. McKendrick's ambition was to found a physiological institute, and in 1903 the Glasgow public provided the money for this to the University. He was at various times examiner in physiology in the universities of London, Oxford, Cambridge, Durham and Aberdeen. He had been Fullerian Professor of Physiology at the Royal Institute of Great Britain, delivered the Thomson Lectures at the Free Church College of Aberdeen, and was one of the lecturers in connection with the Gilchrist Trust. He retired to Stonehaven in 1906 and was for several years a town councillor and Provost of the Burgh, and died in Glasgow on 2 January 1926. He was married in 1867 and had four children.

<u>Honours and Distinctions</u> *

1872   Fellow of the Royal College of Physicians of Edinburgh
1882   Honorary LL.D., Aberdeen University
1884   Fellow of the Royal Society of London
1894   Awarded MakDougall-Brisbane Prize, Royal Society of Edinburgh
1894-99   Vice-president of the Royal Society of Edinburgh

<u>Publications</u> *

Outlines of physiology in its relation to man   (Glasgow, 1878)
General physiology of the nervous system   (London and Glasgow, 1879)
The gases of the blood in relation to some of the problems of respiration (London, 1888)
Elementary human physiology   (London and Edinburgh, 1896)
A text-book of physiology   (Glasgow, 1888)

<u>Biographical Studies</u> *

<u>Proc. R. Soc. Edinb.</u>, <u>46</u>, (Edinburgh, 1927), p.372
<u>Edinb. Med. J.</u> <u>33</u>, (Edinburgh, 1926), pp.176-7
Comrie,J.D., History of Scottish Medicine, <u>2</u>, (London, 1932), pp.608, 655-7,662, 695, 713
Coutts,J., A History of the University of Glasgow, (Glasgow, 1909), pp.469, 474, 579
<u>Br. med. J.</u> <u>33</u>, (London, 1926), p.176

<u>Manuscript Locations</u>

E; L; LR; SaU

Sir Andrew Douglas MACLAGAN (1812–1900)

Professor of Medical Jurisprudence and Public
Health at the University of Edinburgh
(1862–1897)

Elected F.R.S.E.   9 January 1843

<u>Portraits, etc.</u>

Bust: (J. Hutcheson, 1882), University of
   Edinburgh **

Born in Ayr on 17 April 1812, the eldest son of
David Maclagan, Andrew was from a family of
eminent doctors. He received his early
education at the Edinburgh High School and then
went on to study medicine at Edinburgh
University, where he graduated M.D. in 1833.
The title of his thesis was "De Calculis
Biliariis". After graduating, he continued his
studies in London, Paris and Berlin, and then
returned to Edinburgh practising surgery at the Royal Infirmary, and becoming an active member
of the Royal College of Surgeons. In 1845, Maclagan abandoned this to become a lecturer on
materia medica in the extra-mural school of Edinburgh, where he built the foundation for his
great work on toxicology. He held this appointment for nearly eighteen years. In 1862, he became
Professor of Medical Jurisprudence and Public Health at Edinburgh University, remaining in this
post for thirty-five years until he resigned. During this time he acted as the principal medico-
legal adviser to the Crown authorities in Scotland. He was a leading authority on the analysis
of poisons, and almost all the medico-legal cases of this type in Scotland came into his hands.
His contributions to forensic medicine and toxicology were numerous and he constantly urged
responsibility be placed on medical men in giving evidence as skilled witnesses in medico-legal
cases. From 1868 till his death he was Surgeon General of the Royal Company of Archers. He was
also active in the Royal College of Physicians of Edinburgh, honorary member of the
Pharmaceutical Society of Britain, Fellow of the Chemical Society of London and of the Botanical
Society of Edinburgh. As well as holding various presidential positions, Maclagan was probably
best known as a poet and a man of great wit. Many of his lays and poems were on professional or
academic themes, first appearing at some medical or social club dinner and later in magazines.
In 1850, a collection was published under the title "Nugae Canorae Medicae". He died at 28,
Heriot Row, Edinburgh on 5 April 1900 and was buried in the Dean Cemetery, Edinburgh.

<u>Honours and Distinctions</u>

1832-3   President of the Royal Medical Society of Edinburgh
1833   M.D., Edinburgh
1833   Fellow of the Royal College of Surgeons
1859-61   President of the Royal College of Surgeons of Edinburgh
1884   President of the Royal College of Physicians of Edinburgh
1886   Knighted
1890-4   President of the Royal Society of Edinburgh
1891   LL.D., Glasgow University
1897   LL.D., Edinburgh University

<u>Publications</u> *

Notice regarding the composition of James' powder (Edinburgh, 1838)
"On the action of sesquioxide of iron on arsenic", <u>Edinb. Med. J.</u> 1840, (Edinburgh, 1840)
On the constitution of intestinal concretions, (London, 1841)
"On the bebeeru tree of British Guiana", <u>Trans. R. Soc. Edinb.,</u> <u>15,</u> (Edinburgh, 1844)
"On the medicinal quality of Bebeerine", <u>Edinb. Med. Surg. J.</u> <u>63,</u> (Edinburgh, 1845)
"Contributions to toxicology", <u>Edinburgh Monthly Journal,</u> 1848-49, (Edinburgh, 1849)
"Lectures on the general therapeutics", <u>Edinburgh Monthly Journal,</u> 1849-50, (Edinburgh, 1850)
Nugae canorae medicae: lays by the poet laureate of the New Town Dispensary (Edinburgh, 1850)
"On the alkaloids contained in the wood of bebeeru, or greenhart tree",
<u>Trans. R. Soc. Edinb.,</u> <u>25,</u> (Edinburgh, 1869)
Short syllabus of the course of lectures on materia medica (Edinburgh, n.d.)

<u>Biographical Studies</u>

<u>Br. med. J.</u> 1900, i, (London, 1900), pp.935-37
<u>Caledon med. J.</u> 1899-1900, iv, (Glasgow, 1900), p.203
<u>The Lancet,</u> 1900, i, (London, 1900), p.1100
<u>Medical Press and Circular,</u> (London, 1900), pp.lxix, 390
<u>Scottish Medical and Surgical Journal,</u> 1900, iv, (Edinburgh, 1900), pp.451-53
NUC, <u>352,</u> p.495
The London and Provincial Medical Directory, (London, 1859), p.97
Boase, F., Modern English Biography, <u>6,</u> (London, 1965), p.127
Logan-Turner,A., History of the University of Edinburgh, 1883-1933, (London, 1933), p.394

<u>Manuscript Locations</u>

EPH; EU

Robert Craig MACLAGAN   (1839-1919)

Physician

Elected F.R.S.E.   1 February 1869

<u>Portraits, etc.</u>

None known

Born in 1839, the son of Sir Douglas MacLagan,
Robert received his early education at the
Edinburgh High School. He then studied medicine
at Edinburgh University, graduating M.D. in
1860 with the thesis "On Hyoscyamus niger".
Early on in his medical career, he carried out
investigations in connection with the arsenic
eaters of Styria, in Austria. For a while he
practised in Edinburgh, then joined the
University Volunteer Battalion Royal Scots,
where he became a Colonel of the 5th Battalion.
He moved up the ranks to become Honorary
Colonel. He retired in 1909, because of his
deafness, and his interests in folk-lore and
anthropology took over. He wrote many articles
on these subjects. He died on 12 July 1919.

<u>Honours and Distinctions</u>

1866   Fellow of the Royal College of Physicians of Edinburgh

<u>Publications</u> *

Evil Eye in the Western Highlands (London, 1902)
The Perth Incident of 1369 from a folk-lore point of view (Edinburgh and London, 1905)
Religio Scotica: its nature as traceable in Scotic saintly tradition (Edinburgh, 1909)
Our ancestors, Scots, Picts, and Cymry, and what their traditions tell us,
(Edinburgh and London, 1913)

<u>Biographical Studies</u>

Historical sketch and laws of the Royal College of Physicians of Edinburgh, (Edinburgh, 1925),
p.13
BMC, <u>149</u>, 1962, p.288
List of graduates in medicine in the University of Edinburgh from 1705 to 1866,
(Edinburgh, 1867), p.168
<u>Br. med. J.</u>, 1919, ii, (London, 1919), p.93

<u>Manuscript Locations</u>

None known

Sir George Husband Baird MacLEOD   (1828–1892)

Regius Professor of Surgery at the University
of Glasgow

Elected F.R.S.E. 17 January 1870

<u>Portraits, etc.</u>

Portrait: (Unknown), NPG
Photograph: see Comrie, J.D., *History of Scottish
    Medicine*, (London, 1932)
Photograph: Medical Archives Centre, Royal
    Infirmary of Edinburgh **

Born in 1828, the son of Dr. Norman MacLeod of
Glasgow, George studied medicine at Glasgow
University, graduating M.D. in 1853. He then
continued his studies at Paris and Vienna, and
a year after graduation became senior surgeon
to the civil hospital at Smyrna where he
remained for two years. He returned to Glasgow
and was lecturer and professor of surgery in Anderson's College from 1859. In 1869 he succeeded
Joseph Lister as Regius Professor of Surgery at Glasgow University. Although interested in the
development of antiseptic surgery, he was more a practical surgeon than a surgical pathologist.
He taught very large classes in systematic, clinical and operative surgery as well as having an
extensive practice. From 1887 until his death he was a Crown Member of the General Council of
Medical Education. Three years after his death his widow, Lady MacLeod, instituted a memorial
gold medal for the most distinguished student in Surgery Class. He died at Woodside Cresent,
Glasgow on 31 August 1892, and was buried there in Campsie Churchyard.

<u>Honours and Distinctions</u>

1877   Surgeon-in-Ordinary to the Queen in Scotland
1892   Knighted

<u>Publications</u> *

Notes on the surgery of the war in the Crimea, with remarks on the treatment
of gunshot wounds (London, 1858)
Outlines of surgical diagnosis   (New York, 1864)
An exposition of surgical clinical teaching in Glasgow   (London, 1875)
The four apostles of surgery (Hippocrates, Galen, Pare, Hunter): an historical sketch
(Glasgow, 1877)
Inquiries and diseases of the neck   (New York, 1884)
A trip to Snowland (London, 1891)

<u>Biographical Studies</u>

Boase, F., *Modern English Biography*, <u>2</u>, (London, 1965), p.655
Coutts, J., *History of the University of Glasgow*, (Glasgow, 1909), pp.584-5
Comrie, J.D., *History of Scottish Medicine*, <u>2</u>, (London, 1932), pp.639-640, 661
<u>Proc. R. Soc. Edinb.</u>, <u>19</u>, (Edinburgh, 1892), pp.xl-liii

<u>Manuscript Locations</u>

EU; GC

Robert Bowes MALCOLM (1808-1894)

Physician

Elected F.R.S.E.   7 December 1857

<u>Portraits, etc.</u>

None known

Robert Bowes Malcolm, an Englishman by birth, was the son of Major John Malcolm, who was in the service of the East India Company. Robert was educated in the North of England and later at Edinburgh University where he graduated M.D. in 1831, with the thesis 'De Menorrhagia'. Nine years later he was made a Fellow of the Royal College of Physicians of Edinburgh. His marriage to the daughter of the eminent obstetric physician, Dr. John Thatcher, proved useful to his career. Aided by this, and with his own specialised knowledge of obstetrics, Dr. Malcolm soon climbed to the top of his profession in Edinburgh. His professional distinction was achieved despite the fact he did not teach and does not appear to have been interested in regularly publishing the results of his work, although two pieces of his work appeared in the 'Edinburgh Medical Journal'. His professional relationship with Lady Fettes helped his candidacy for the post of physician to Fettes College, a post which he held for many years with the assistance of Dr. J.D. Gillespie. He was one of thirty doctors concerned with the founding of the Edinburgh Obstetrical Society in 1840. He died in 1894, his wife having died many years before him. They had three children; one son, who was surgeon-major of the 9th Lancers, and two daughters.

<u>Honours and Distinctions</u> *

1840	Fellow of the Royal College of Physicians of Edinburgh
n.d.	External Member of the Royal Medical Society of Edinburgh
n.d.	President of the Royal Physical Society of Edinburgh
1843 & 1849-53	Vice-President of the Obstetric Society of Edinburgh
1856-57	President of the Harveian Society of Edinburgh

<u>Publications</u>

"On the Rise of the Edinburgh School of Midwifery - Annual address for 1856", <u>Edinb. Med. J.</u> (Edinburgh, 1856)
"Case illustrative of spontaneous evolution", <u>Edinb. Med. Surg. J.</u>, (Edinburgh, 1834)
"Rupture of uterus with hydrocephalus of the infant", <u>London and Edinburgh Monthly Journal of Medical Science</u>, (Edinburgh, n.d.)

<u>Biographical Studies</u>

<u>Edinb. Med. J.</u> 40, (Edinburgh, 1894), pp.90-91
Craig,W.S. McRae, History of the Royal College of Physicians of Edinburgh, (Oxford, 1976), pp.297, 956
Wemyss,H.L. Watson, A record of the Edinburgh Harveian Society, (Edinburgh, 1833), p.54
Medical Directory for 1885, (London, 1885), p.1065

<u>Manuscript Locations</u>

EU

Sir Arthur MITCHELL   (1826-1909)

Commissioner in Lunacy; Antiquary

Elected F.R.S.E.   19 February 1866

<u>Portraits, etc</u>

Portrait: (N. Macbeth, R.S.A., 1880), NPG **
Portrait: (Sir George Reid, P.R.S.A., 1896), SNPG

Born at Elgin on 19 January 1826, the son of George Mitchell, C.E., Sir Arthur received his early education at Elgin Academy. He then studied Arts at Aberdeen University, graduating M.A. in 1845, but changed to medicine, and graduated M.D. three years later. He continued his studies in Paris, Berlin and Vienna. From the beginning of his career he was involved in research into mental illness and was concerned with improving the treatment of the mentally ill. He was Morison Lecturer on Mental Diseases to the Royal College of Surgeons (1867-71). From 1870 to 1895 he was Commissioner in Lunacy for Scotland while in 1880, he was a member of the Commission on Criminal Lunacy (England). Mitchell also acted as Chairman of the Commission on Lunacy (Ireland), 1888-91. Apart from his medical work, Sir Arthur was also actively engaged in his archaeological and meteorological interests. From 1876-78 he was Rhind Lecturer in Archaeology and was made a Fellow of the Society of Antiquaries of Scotland and of the Meteorology Society. He died at 34 Drummond Place, Edinburgh on 12 October 1909 and was buried at Rosebank Cemetery.

<u>Honours and Distinctions</u>

1876   Honorary LL.D., Aberdeen University
1886   Companion of the Order of the Bath
1887   Knight Commander of the Bath
1908   President of the Royal Meteorology Society
n.d.   Council Member of the Scottish Meteorology Society

<u>Publications</u>

On various superstitions in the N.W. Highlands and Islands of Scotland especially in relation to lunacy (Edinburgh, 1862)
On the popular weather prognostics of Scotland (Edinburgh, 1863)
The insane in private dwellings (Edinburgh, 1864)
The past in the present. What is civilization? (Edinburgh, 1880)
A list of travels and tours in Scotland 1220 to 1900 (Edinburgh, 1902)
About dreaming, laughing and blushing (Edinburgh, 1905)
Editor of 'Geographical Collections relating to Scotland made by Walter Macfarlane' (Edinburgh, 1906-1908)

<u>Biographical Studies</u>

DNB, (London, 1912), second supplement
Who was who 1897-1916, <u>1</u>, 5th ed., (London, 1967)
Eddington, A., Edinburgh and the Lothians at the opening of the twentieth century, (Edinburgh, 1904), p.232

<u>Manuscript Locations</u>

EU; L; SaU

John MOIR   (1808–1899)

Professor of Midwifery at the University of
Edinburgh

Elected F.R.S.E.   6 March 1865

<u>Portraits, etc.</u>

Portrait: (C. MacPherson, 1869), RCPE **

John Moir was born in a French fortress at
Verdun on 6 April 1808, the son of James Moir,
a naval surgeon who was taken prisoner during
the Napoleonic Wars. John received his early
education at the Edinburgh High School, when
his parents returned to the city. John then
studied medicine at the University, graduating
M.D. in 1828 with his thesis 'De Febre
Intermittence'. He joined his father in his
practice in Edinburgh and became assistant to
Professor Hamilton in the University, and often
lectured for him. He was Assistant Physician at the Edinburgh Maternity Hospital, until becoming
Professor of Midwifery at the University of Edinburgh. During this time he was also a private
lecturer on the subject, and had a large practice at the same time. He was an intimate friend of
James Young Simpson, and lived to become the oldest Fellow of the Royal College of Physicians in
Edinburgh. Moir was consulting physician of the Royal Maternity Hospital for many years, and was
one of the founder members of the Edinburgh Medical Missionary Society. He died in Edinburgh on
14 May 1899, and was buried there in the New Calton Burying Ground.

<u>Honours and Distinctions</u>

1837   Fellow of the Royal College of Physicians of Edinburgh
1854-5   Vice-president of the Obstetrical Society of Edinburgh
1858-9   President of the Obstetrical Society of Edinburgh
1865-7   President of the Medico-Chirurgical Society of Edinburgh
1867-8   President of the Royal College of Physicians of Edinburgh

<u>Publications</u>

On retroflexion of the unimpregnated uterus; with cases illustrative of its causes and of a new
mode of treatment (Edinburgh, 1860)
Valedictory address on retiring from the President's Chair of the Medico-Chirurgical Society of
Edinburgh, 1867, (Edinburgh, 1868)
Induction of premature labour   (Edinburgh, 1898)

<u>Biographical Studies</u>

<u>Transactions of the Edinburgh Obstetrical Society,</u> <u>25,</u> 1899-1900, p.3
<u>Edinb. Med. J. 5,</u> (Edinburgh, 1899), p.638
<u>Proc. R. Soc. Edinb.,</u> <u>23,</u> (Edinburgh, 1901), p.9
<u>The Times,</u> 15 May 1899, p.11
Anderson, W.P., Silences that Speak, (Edinburgh, 1931), p.650
Boase, F., Modern English Biography, <u>6,</u> (London, 1965), p.225

<u>Manuscript Locations</u>

EU

Charles MOREHEAD   (1807-1882)

Physician in the Bombay Medical Service

Elected F.R.S.E.   15 January 1866

<u>Portraits, etc.</u>

Drawing: see H. Haines, Memorial of the life and
    work of Charles Morehead (London, 1884) **

Born in Edinburgh in 1807, the second son of
Robert   Morehead,   rector   of   Easington,
Yorkshire, Charles received his early education
at the Edinburgh Royal High School. After
attending   a   short   course   of   lectures   at
Glasgow, he studied medicine at the University
of Edinburgh, graduating M.D. with the thesis
"De   sputis   quatenus   ad   Morbos   Pectoris
dignoscendum conferunt", in 1828. He studied
in Paris for a while under Pierre Louis, before
going to India and becoming a member of the
Bombay Medical Service in 1829. Two years later he was transferred to the staff of Sir Robert
Grant, governor of Bombay and in 1835 he became secretary of the Bombay Medical and Physical
Society, a society he had originated for the advancement of medical science. He then became a
member of the staff of the European and Native General Hospitals of Bombay in 1838, and two
years after that he was appointed secretary to the Board of Education, where he had the
opportunity to demonstrate his devotion to the late Governor Grant's ideals of establishing an
Indian medical profession among the natives of India. To promote this the Grant Medical College
in Bombay was opened in 1845, and Morehead was its first Principal, holding this post for
fifteen years. After his death, 'The Charles Morehead Scholarship in Chemical Medicine' was
established by the University of Bombay in memory of Morehead, to be awarded annually to the
most promising graduate in medicine of Grant College. He retired to England in 1862, where he
was regarded as an authority on tropical diseases and medical education. He died on 24 August
1882 at Wilton Castle, Yorkshire.

<u>Honours and Distinctions</u>

1860   Fellow of the Royal College of Physicians of London
1861   Honorary Surgeon to the Queen
1881   Companion of the Order of the Indian Empire

<u>Publications</u> *

Notes on Bright's disease of the kidneys (Bombay, 1850)
An introductory lecture delivered in the Grant Medical College at Bombay, 1850 (Bombay, 1850)
An introductory lecture delivered in the Grant Medical College at Bombay, 1853 (Bombay, 1853)
Clinical researches on diseases in India (London, 1856)
Notes on the prevention and treatment of cholera (Edinburgh, 1866)
Memorials of the life and writings of Robert Morehead (Edinburgh, 1875)

<u>Biographical Studies</u> *

Haines, Hermann A., (ed). Memorial of the life and work of Charles Morehead,
(Privately printed, 1884)
<u>Proc. R. Soc. Edinb.</u>, <u>14</u>, (Edinburgh, 1887), pp.41-3
Gray,J., History of the Royal Medical Society of Edinburgh, 1737-1937, (Edinburgh, 1952),
pp.170-1
Munk,W., Roll of the Royal College of Physicians of London, <u>4</u>, (London, 1955), pp.129-30

<u>Manuscript Locations</u>

None known

John Ivor MURRAY   (1824–1903)

Physician

Elected F.R.S.E.   2 February 1857

<u>Portraits, etc.</u>

None known

Born at Lasswade, Edinburgh, in 1824, the
second son of John Murray, W.S., John received
his early education at the Lycee Sain Louis,
in Paris. He then studied medicine at Edinburgh
University. In 1844, he gained as a prize a
commission as assistant surgeon in the army. He
then proceeded to China, travelling through
Borneo to Canton, where he was in charge of the
hospital during the outbreak and attack
on European factories in 1846. He later moved
from Canton to Shanghai, where he set up a
large practice. In 1852 at his own cost, he
built the first hospital for Europeans in the North of Hong Kong. At the beginning of the
Crimean War, Murray travelled to Sebastapol, being appointed second class staff surgeon, and
acting second-in-charge of the General Hospital at Balaclava. He received the Turkish and
Crimean Medals, and the Sebastapol Clasp for his services. In 1856, he returned to Britain where
he took his M.D. at his old university. He returned to Hong Kong, where he took up the position
of Colonial Surgeon, a post he held from 1858 to 1872. He was also Inspector of Hospitals in
1868. In 1872, he returned again to England setting up a large practice in Scarborough 3 years
later. He was the Honorary Consulting Physician to the Sea-Bathing Infirmary and the Cottage
Hospital. He retired in 1890, and returned to London. Soon afterwards, he became the junior
warden of the Grand Lodge of the North and East Ridings in Yorkshire. He retired again in 1900,
and spent the last years of his life in London, where he was elected President of the British
Balneological Society. He died at Addison Mansions on 24 July 1903.

<u>Honours and Distinctions</u>

1856   Fellow of the Royal College of Surgeons of Edinburgh
1867   Order of Merit from the King of Sweden
1868   Order of Merit from the King of Italy
1868   Inspector of Hospitals, Hong Kong
1895   Royal Medal, Royal Society of London
1900   President of the Hunterian Medical Society, Edinburgh
1900   President of the British Balneological Society

<u>Publications</u>

"Note of the result of an analysis of a portion of the bread with which A-lum was accused of
poisoning European residents at Hong Kong", <u>Edinb. Med. J.</u> <u>4</u>, 1858, (Edinburgh, 1859)

<u>Biographical Studies</u>

The Record of the Royal Society, 1897, (London, 1897), p.217
<u>Br. med. J.</u> <u>2</u>, (London, 1903) pp.339–40
<u>The Lancet</u>, <u>2</u>, (1903), p.504
The London and Provincial Medical Directory, 1859, (London, 1859), p.103

<u>Manuscript Locations</u>

None known

Robert NAYSMITH (1792-1870)

Surgeon and Dentist

Elected F.R.S.E.   22 February 1842

<u>Portraits, etc.</u>

None known

Born in Edinburgh in 1792, Naysmith was educated at the city's High School. At the age of fifteen he began studying medicine at Edinburgh University, where he proved to be a talented pupil. He often assisted Wardop, Liston, Syme and Fergusson when they operated. He decided to go to London for a time, where he studied dentistry. Mr Naysmith was of the opinion that a practitioner should be fully qualified in all fields, and he was not only a skilful dentist but also an accomplished surgeon. He soon gained public confidence, and eventually had a larger practice in Edinburgh than any dentist before him. He was appointed Surgeon-Dentist to the Royal Household and held this position during the reigns of three successive sovereigns - George IV, William IV, and Queen Victoria. He supported the first establishment of the Dental Dispensary in Edinburgh, and continued as one of its medical staff. For some time before his death he had been in feeble health. He died in his house in Charlotte Square on 12 May 1870, at the age of seventy-eight.

<u>Honours and Distinctions</u>

1823   Fellow of the Royal College of Surgeons
1857   Vice-President Odontological Society of London

<u>Publications</u>

"Physiology and Pathology of the teeth", <u>London and Edinburgh Journal of Medical Science</u> (London, 1843)

<u>Biographical Studies</u>

<u>Proc. R. Soc. Edinb.</u>, <u>7</u>, (Edinburgh, 1872), pp.245-6
<u>Edinb. Med. J.</u>, <u>15</u>, (2), (Edinburgh, 1870), p.1149

<u>Manuscript Locations</u>

E

Patrick Small Keir NEWBIGGING (1813-1864)

Physician

Elected F.R.S.E.   17 April 1848

<u>Portraits, etc.</u>

None known

Born in Edinburgh on 2 November 1813, the
youngest son of Sir William Newbigging, M.D.,
Patrick studied medicine at Edinburgh
University and graduated M.D. in 1834, with the
thesis "On the causes of the impulse and sounds
of the heart". After further study at various
European universities, he returned to Edinburgh
and was elected one of the Medical Officers of
the New Town Dispensary. In 1842 he published a
translation of Barth and Roger's "Treatise on
Ausculation", and four years later, on his
father's death, he succeeded to his practice.
He was also at various times a surgeon at John Watson's Institute and Chauvin's Hospital, as
well as being medical referee to the Life Association Insurance Company of Scotland, one of the
original physicians to the Sick Children's Hospital, and an examiner of the Royal College of
Surgeons. He died in Edinburgh on 10 January 1864, and was buried there in the Dean Cemetery.

<u>Honours and Distinctions</u> *

c.1834   President of the Royal Medical Society, Edinburgh
1834   Fellow of the Royal College of Surgeons of Edinburgh
1861-63   President of the Royal College of Surgeons of Edinburgh
1862   President of the Royal Scottish Society of Arts, Edinburgh

<u>Publications</u> *

Probationary essay on spina bifida (Edinburgh, 1834)
Inaugural essay on the impulse and sounds attending the action of the heart in its normal state
(Edinburgh, 1834)
On certain circumstances affecting the colour of blood during coagulation (Edinburgh, 1839)
Case of tetanus, in which recovery took place (n.p., 1845)
Observations on the therapeutic action of croton oil in nervous disorders,
<u>Edinb. Med. Surg. J.</u> <u>55</u>, (Edinburgh, 1841)
'Statistics at Edinburgh Lock Hospital during 1841, 1842, bearing in particular on the non-
mercurial treatment of the venereal disease', <u>London and Edinburgh Monthly Journal of Medical
Science</u>, <u>3</u>, 1843; <u>7</u>, 1847, (Edinburgh, 1843, 1847)
Notes on scarlatina (Edinburgh, 1849)
Account of illness and death of W.P. Alison (Edinburgh, 1860)

<u>Biographical Studies</u>

<u>Edinb. Med. J.</u> <u>9</u>, (Edinburgh, 1864), p.773
Royal College of Physicians of London, Library Catalogue, (London, 1912), p.882
Biographisches Lexikon der Hervorragenden Aerzte, <u>4</u>, (Berlin, 1929), pp.353-4
<u>Trans. R. Scott. Soc. Arts.</u> <u>5</u>, (Edinburgh, 1861), pp.2, 39, 52
Rogers, R.C., Monumental Inscriptions in Scotland, <u>1</u>, (London, 1871), p.143
List of the graduates in medicine in the University of Edinburgh, from 1705 to 1866,
(Edinburgh, 1867), p.102

<u>Manuscript Locations</u>

EU

Samuel Alexander PAGAN (-)

Physician

Elected F.R.S.E. 5 January 1846

<u>Portraits, etc.</u>

None known

Nothing is known of Pagan's early life. He studied medicine at Edinburgh University, graduating M.D. in 1824 with the thesis 'De Rheumatismo'. Pagan seems to have established his career in Edinburgh, which included the positions of Consulting Medical Officer with the New Town Dispensary and that of Ordinary Manager of the Royal Infirmary and Royal Lunatic Asylum. He was elected President of the Royal College of Surgeons of Edinburgh in 1846 and was also a member of the Edinburgh Harveian Society, becoming President in 1849. Latterly Pagan lived in Melville Street, Edinburgh.

<u>Honours</u> and <u>Distinctions</u>

1829   Fellow of the Royal College of Surgeons of Edinburgh
1846-7   President of the Royal College of Surgeons of Edinburgh
1849   President of the Edinburgh Harveian Society

<u>Publications</u>

"Vesico-Vaginal Fistula", <u>London and Edinburgh Monthly Journal, 1842</u>, (London, 1842)
A probationary essay on scrofula, (Edinburgh, 1829)

<u>Biographical Studies</u>

<u>Proc. R. Soc. Edinb.</u>, <u>6</u>, (Edinburgh, 1869), p.36
Royal College of Surgeons of Edinburgh, List of Fellows at 31st December 1969, (Edinburgh, [1969]), p.xi
Wemyss, H.L.W., A Record of the Edinburgh Harveian Society (Edinburgh, 1833), p.40
List of graduates in medicine in the University of Edinburgh, from 1705 to 1866, (Edinburgh, 1867), p.73

<u>Manuscript Locations</u>

None known

Alexander PEDDIE (1810–1907)

Physician

Elected F.R.S.E. 5 January 1863

<u>Portraits, etc.</u>

Portrait: (B. Peddie, 1879), RCPE **

Born in 1810, the son of a minister in Edinburgh, Alexander received his education at the High School. He then began a career in banking but gave this up after a few years and changed to studying medicine at Edinburgh in 1830, graduating five years later with the thesis 'On Erysipelas'. He did further studies at Paris, returning to Edinburgh where he was the first to introduce the stethescope. Alexander set up a practice, and along with Dr. Brown and Dr. Cornwall, he took charge of the hospital at Minto House. He was one of several founders of the Royal Hospital for Sick Children. He contributed many papers to various journals, particularly the 'Edinburgh Medical Journal'. Later in his life, he took up his favourite hobbies of music, painting and fishing. He died on 19 January 1907, after an attack of pneumonia, and was buried at Warriston Cemetery.

<u>Honours and Distinctions</u>

1845   Fellow of the Royal College of Physicians of Edinburgh
1877   President of the Royal College of Physicians of Edinburgh
1890   Member of the Harveian Society

<u>Publications</u> *

Cases of dropsy and gangrene occurring in a family who had subsisted for some time on unwholesome potatoes (Edinburgh, 1833)
On the mammary secretion (n.p., 1848)
On the pathology of delirium tremens, and its treatment without stimulants or opiates (Edinburgh, 1854)
The necessity of legalised arrangements for treatment of dipsomania; or the drinking insanity (Edinburgh, 1858)
On the relations of drink and insanity, with reference to proposed legalisation for habitual drunkards (Lewes, [1876])
Recollections of Dr. John Brown, author of 'Rab and his friends', etc., with a selection from his correspondence, by A. Pirrie (London, New York, 1893; Edinburgh 1894)

<u>Biographical Studies</u>

<u>Edinb. Med. J.</u>, <u>21</u>, (Edinburgh, 1907), pp.97–100
Royal College of Physicians of London, Library Catalogue, (London 1912), p.938
List of graduates in medicine in the University of Edinburgh, from 1705 to 1866, (Edinburgh, 1867), p.105
Historical sketch and laws of the Royal College of Physicians of Edinburgh, (Edinburgh 1925), pp.10, 25
<u>Br. med. J.</u>, <u>1</u>, (London, 1907), pp.291–293

<u>Manuscript Locations</u>

None known

James Bell PETTIGREW (1834–1908)

Chandos Professor of Medicine and Anatomy at
the University of St. Andrews

Elected F.R.S.E. 3 March 1873

<u>Portraits, etc.</u>

Photograph: see Comrie, J.D., History of Scottish
    Medicine, (London, 1932) **
Portrait: (W.W. Ouless, c.1902), location unknown
Photograph: see Leyland, J., Contemporary medical
    men (Leicester, 1888)

Born at Roxhill, Lanarkshire, on 26 May 1834,
son of Robert Pettigrew, James received his
early education at the Free West Academy of
Airdrie. He then studied Arts at Glasgow
University for five years, before changing to
medicine at Edinburgh University. He was a
distinguished student, and was a gold medallist
in both Anatomy and Medical Jurisprudence. In 1860 he was Croonian Lecturer to the Royal Society
of London, and when graduating M.D. the following year, the Senatus of Edinburgh University
conferred on him a gold medal for his thesis "On the ganglia and nerves of the heart...". In
1862 he became assistant curator of the Hunterian Museum at the Royal College of Surgeons of
England, and seven years later, was appointed curator of the Museum of the Royal College of
Surgeons, Edinburgh, having spent the previous two years in Ireland studying the flights of
insects, birds and bats. At the Royal College of Surgeons, he also lectured on physiology and in
1873 was examiner in the subject for both the College of Surgeons and Physicians. In 1875 he
became Chandos Professor of Medicine and Anatomy at the University of St. Andrews, as well as
Dean of the Medical Faculty. From 1883, for a period of four years, he was examiner in Anatomy
in the University of Glasgow. In 1877 he was elected by the Universities of Glasgow and St.
Andrews as a representative on the General Medical Council. His studies on the motion of flight
in animals were an important contribution in the development of aeronautics. He also contributed
to the study of the valves, and of the arrangement of muscular tissue in the heart. He died at
his home in St. Andrews on 30 January 1908.

<u>Honours and Distinctions</u> *

1860   President of the Royal Medical Society
1872   Fellow of the Royal Society of London
1873   Fellow of the Royal College of Physicians
1874   Awarded the Godard Prize by the French Academie des Sciences
1874   Made Laureate of the Institute of France

<u>Publications</u> *

"On the relations, structure and function of the valves of the vascular system in Vertebrata",
<u>Trans. R. Soc. Edinb.,</u> <u>23</u>, (Edinburgh, 1864)
"On the various modes of flight in relation to aeronautics", <u>Proc. R. Instn. Gt. Br.,</u> Annual
Report (1867), (Washington, 1868)
On the relation of plants and animals to inorganic matter, and on the interaction of the vital
and physical forces (Edinburgh, 1873)

<u>Biographical Studies</u>

Who Was Who, 1897-1915, 5th ed., (London, 1967)
Comrie, J.D., History of Scottish Medicine, <u>2</u>, (London, 1932), pp.578-579

<u>Manuscript Locations</u>

SaU

William PIRRIE   (1807–1882)

Regius Professor of Surgery, Marischal College,
Aberdeen

Elected F.R.S.E.   2 April 1849

<u>Portraits, etc.</u>

Photograph: see Comrie, J.D., History of Scottish
    Medicine, (London, 1932) **

Born near Huntly, Aberdeenshire in 1807, the
son of George Pirrie, a farmer, William was
educated at Gartly Parish School. He then
studied at Marischal College and the University
of Aberdeen, graduating M.A. in 1825, before
going to Edinburgh to study medicine. He
graduted M.D. there four years later, with the
thesis "De Vitiis Tali" and then went to Paris
to study surgery under Baron Dupuytren. On his
return to Aberdeen in 1830, he became a
lecturer on anatomy and physiology in King's and Marischal Colleges, and when the two separated
in 1839, he was appointed the first Regius Professor of Surgery in Marischal College. When the
colleges united again in 1860, he continued as Professor of Surgery. Altogether he held office
for fifty-two years, during which time he enhanced the reputation of the Aberdeen Medical
School. He was considered pre-eminent in his profession in the North of Scotland. He built up a
large collection of old and new surgical instruments, and did much to establish medical
bursaries at Aberdeen University. It was through his influence that a Chair of Pathological
Anatomy was founded at the University in 1882. He was an enthusiatic promoter of the method of
arresting haemorrhage in surgery by acupressure. He died on 21 November 1882 at 253, Union
Street, Aberdeen.

<u>Honours and Distinctions</u>

n.d.   President of the North of Scotland Medical Association
1875   Honorary LL.D., University of Edinburgh
1877   Surgeon to H.R.H The Prince of Wales in Scotland

<u>Publications</u>*

"Remarks on the modes of proceeding in regard to the hernial sac in the operation for
strangulated hernia" <u>Edinburgh Monthly Journal of Medical Science,</u> (Edinburgh, 1848)
The principles and practice of surgery (London, 1852)
Observations on favus (Aberdeen, 1861)
On hay asthma, and the affection termed hay fever (London, 1867)
Acupressure: an excellent method of arresting surgical haemorrhage and of accelerating the
healing of wounds (London, 1867)

<u>Biographical Studies</u>*

DNB, (London, 1968)
Comrie, J.D., History of Scottish Medicine, <u>2</u>, (London, 1932), pp.393, 550-4, 560, 567, 737
<u>Edinb. Med. J.</u> 28, (2), (Edinburgh, 1883), pp.662-8

<u>Manuscript Locations</u>

E; L

Sir John RICHARDSON (1787-1865)

Surgeon, Arctic Explorer and Naturalist

Elected F.R.S.E. 1855

<u>Portraits, etc.</u>

Print: see Comrie,J.D., History of Scottish Medicine, (London, 1932) **

Born at Nith Place, Dumfries, on 5 November 1787, the eldest child of Provost Gabriel Richardson, Sir John received his early education at Dumfries Grammar School, and was then apprenticed for a year to his uncle, James Mundell, surgeon in Dumfries. In 1801 he went to Edinburgh University to study medicine, and after returning to his home town for two years, he came back to Edinburgh in 1806, qualified as a member of the Edinburgh Royal College of Surgeons, and joined the Royal Navy as assistant surgeon. During his naval career he was present at the blockade of the Russian Fleet in the Tagus. He was sent to Madeira and Cape Coast Castle as a surgeon and served on convoy duty to Spain and Quebec. He returned to Edinburgh to complete his medical studies, concentrating on botany and mineralogy, and graduating M.D. in 1816. He began practising as a physician in Leith, but had little success. Two years later he was appointed surgeon and naturalist on Franklin's Polar expedition, which lasted four years, covering 5,550 miles while in America. In 1822 he became surgeon to the Chatham division of marines. When three years later he was asked to join Franklin's second expedition to the mouth of the Mackenzie, he was released from this post. After a short while he and Franklin separated, Richardson setting off with his own crew to explore 900 miles of coast. This he did in less than a month, travelling almost 2,000 miles. After this he was physician in England for some years, becoming Physician to the Royal Hospital at Haslar in 1838 and Inspector of Hospitals in 1840. Eight years later he went on a search expedition for Sir John Franklin which lasted a year. He retired from service shortly after this, spending his last years at Lancrigg, Grasmere, writing, acting as a magistrate, and giving medical aid to the poor. He died there on 5 June 1865 and was buried at Grasmere Churchyard.

<u>Honours and Distinctions</u>

1825   Fellow of the Royal Society of London
1826   Royal Medal, Royal Society of London
1846   Knighted
1850   Companion of the Bath
1855   Honorary Fellow of the Royal Society of Edinburgh
1857   Honorary LL.D., University of Dublin

<u>Publications</u> *

Arctic searching expedition: a journal of a boat voyage through Rupert's Land and the Arctic Sea, in search of Sir John Franklin (London, 1851)
The zoology of the voyage of H.M.S. Herald; Fossil remarks by Sir John Richardson (London, 1852-54)
Fauna Boreali-Americana; or the zoology of the Northern parts of British Americana (3 vols) (London, 1829-37)
Reports on the Ichthyology of the Seas of China and Japan (London, 1846)
The Polar Regions (1861)

<u>Biographical Studies</u> *

DNB, (London, 1968)
<u>Proc. Linn. Soc.</u>, 1865-66, (London, 1866)
MacLaith, John, Life of Sir John Richardson, 8 vols. (n.p., 1868)
Rolleston, Sir H.D., "Sir John Richardson, C.B., M.D., LL.D., F.R.S.",
<u>J. Roy. Nav. Med. Serv.</u>, 10, (1924), p.161
Comrie, J.D., History of Scottish Medicine, 2, (London, 1932), pp.723, 744, 745
Boase, F., Modern English Biography, 3, (London, 1965), p.150
Curvey, Mary F., Johnson, Robert E., "A bibliography of Sir John Richardson. Printed books",
<u>J. Soc. Biblphy. nat. Hist.</u>, 5, (1969), pp.202-17
Huntley, M.A., Johnson, R.E., (etc.), "A bibliography of Sir John Richardson - articles in learned journals", <u>J. Soc. Biblphy. nat. Hist.</u>, 6, (1971), pp.98-117

<u>Manuscript Locations</u>

BM(IH); CSR; E; EU; KB; L; LGS; LLS; SaU;

Douglas Argyll ROBERTSON   (1837-1909)

Ophthalmic Surgeon

Elected F.R.S.E.   1 April 1872

<u>Portraits, etc.</u>

Portrait: (Sir George Reid), RCPE
Photograph: see Comrie,J.D., History of Scottish
    Medicine, (London, 1932) **

Robertson was born in Edinburgh in 1837, the son of Dr. John Argyll Robertson, surgeon and one of the founders of the Edinburgh Eye Dispensary. He was educated at the Edinburgh Institution, Neuwied in Germany and the Universities of Edinburgh and St. Andrews, where he graduated M.D. in 1857. In the same year he became house-surgeon at the Edinburgh Royal Infirmary, and then went to Berlin to study ophthalmic surgery under Von Graefe. On his return to Edinburgh, he was appointed assistant to Professor John Hughes Bennett, and set up the first course in practical physiology in the University. In 1862 he became a Fellow of the Royal College of Surgeons of Edinburgh, and published his paper on the Calabar bean, proving that physostigmine led to constriction of the pupil of the eye, thereby providing a satisfactory miotic. This discovery attracted universal attention. He was the first in the U.K to take up ophthalmic surgery independently, and introduced new methods of procedure, such as trepanning the sclerotic for relief of glaucoma. Today his name is remembered by the pupil reaction known as the "Argyll Robertson Pupil". In 1867 he was appointed assistant ophthalmic surgeon to the Royal Institution, London, under Dr. William Walker. On the latter's retirement in 1897, Robertson succeeded him as consulting surgeon. He was also Surgeon Oculist in Scotland to Queen Victoria and to King Edward VII. He retired in 1904, and spent his last years in Mon Plaisir, St. Aubyn's in Jersey. On his third visit to India in 1909, he died at Gondal on 3 January, and was cremated on the banks of the River Gondli.

<u>Honours and Distinctions</u>

1886-87   President of the Royal College of Physicians of Edinburgh
1893   President of the Ophthalmic Society of Great Britain
1894   Presided over the International Ophthalmological Congress, Edinburgh
1896   Presided over the Edinburgh Medico-Chirurgical Society
1896   Honorary LL.D., University of Edinburgh

<u>Publications</u> *

A new operation for ectropion   (Edinburgh, 1883)
Note on the further history of the case of filaria loa previously
reported to the Society (London, 1897)
On an interesting series of eye symptoms in the case of spinal disease, with remarks on the action of belladonna on the iris, etc .... four cases of spinal myosis; with remarks on the action of light on the pupil .....On the Calabar bean as a new agent in ophthalmic medicine (Baltimore, 1937)

<u>Biographical Studies</u> *

Comrie, J.D. History of Scottish Medicine, <u>2</u>, (London, 1932), pp.608, 627, 631, 701, 708-9, 713
Who Was Who 1897-1915, 5th ed., (London, 1967)
Guthrie, Douglas J., A History of Medicine (London, 1958), p.376

<u>Manuscript Locations</u>

EU.

William ROBERTSON   (1818–1882)

Physician

Elected F.R.S.E.   16 January 1860

<u>Portraits, etc.</u>

None known

William was born in Edinburgh on 8 January 1818, the eldest son of George Robertson, Keeper of the Records in H.M. General Register House. He received his early education at the Edinburgh Academy and then studied medicine at Edinburgh University, and at Paris, Berlin and Vienna. He graduated M.D. at Edinburgh in 1839 with his thesis "Enlargement of the heart", and was appointed Medical Officer to the New Town Dispensary. Six years later he moved to the Royal Infirmary as junior ordinary physician, rising to senior ordinary physician in 1853, before resigning two years later. In the Crimean War he was inspecting physician to the British Civil Hospital at Renkioi for eighteen months. He was at one time, editor of the 'Edinburgh Monthly Journal of Medical Science', contributing several papers to it. In 1874 he became Superintendent of the Statistical Department in the General Registry Office, having previously been Medical Registrar for Scotland. He was experienced in life assurance matters, and was medical referee to the Guardian, and the Scottish Equitable Societies, and finally, the Scottish Widows' Fund. He had a high reputation as a mathematician and statistician. He died at his home in Albany Street, Edinburgh, on 25 August 1882, and was buried in Warriston Cemetery on 29 August.

<u>Honours and Distinctions</u>

1843   Fellow of the Royal College of Physicians of Edinburgh

<u>Publications</u>

Report on the causes of death among the assured in the Scottish Equitable Life Assurance Society, from 1831 to 1864 (Edinburgh, 1865)

<u>Biographical Studies</u>

<u>Edinb. Med. J.</u>, <u>28</u>, (1), (Edinburgh, 1882), pp.382–4
<u>Proc. R. Soc. Edinb.</u>, <u>14</u>, (Edinburgh, 1887), pp.22–4
Comrie, J.D.,   History of Scottish Medicine, <u>2</u>, (London, 1932), p.628

<u>Manuscript Locations</u>

EU; L; SaU

William RUTHERFORD (1839-1899)

Professor of the Institutes of Medicine at the
University of Edinburgh

Elected F.R.S.E.   18 January 1869

<u>Portraits, etc.</u>

Bust: (J. Hutchinson, RSA, 1900), University of
    Edinburgh
Photograph: see Comrie, J.D., History of Scottish
    Medicine, (London, 1932) **
Photograph: see MSS. Cat., EUL

Born at Ancrum Craig, Roxburghshire, on 20
April 1839, the youngest son of Thomas
Rutherford, William received his early
education at the district grammar school. He
then studied medicine at Edinburgh University
and graduated M.D. in 1863, winning a gold
medal for his thesis. For a short while he was
house-physician at the Edinburgh Royal Infirmary, and assistant demonstrator at Surgeons' Hall,
before travelling through Germany and France to study experimental physiology. In 1864, he spent
a year studying electrical physiology under Professor Du Bois Reymond in Berlin. He travelled
all over Europe before returning to Edinburgh in 1865, when he was appointed assistant to John
Hughes Bennett. Under his influence, Rutherford established his reputation as a practical
teacher, which led to his being offered the post of Professor of Physiology at King's College,
London, in 1869. Then he filled the post of Fullerian Professor of Physiology at the Royal
Institute of London in 1871. On his return to Edinburgh in 1874, he became Professor of
Physiology at the University, continuing in this post until his death. As a physiologist
Rutherford was interested in the recondite problems of electro-physiology, the physiological
action of drugs on the secretion of the bile, and in later life investigated the structure of
the striated muscle and the mechanism of the senses. He died at 14, Douglas Crescent, Edinburgh
on 21 February 1899 and was buried at Ancrum Churchyard.

<u>Honours and Distinctions</u>

1865   Member of the Royal College of Surgeons of England
1876   Fellow of the Royal Society of London

<u>Publications</u>

An introduction to study of medicine (Edinburgh, 1871)
An experimental research on the physiological actions of drugs on the secretion of bile
(Edinburgh, 1879)
A text book of physiology (Edinburgh, 1880)
On the conditions that influence the attainment of the physiology ideal (Edinburgh, 1890)
On the service rendered to mankind by medical science (Edinburgh, 1891)
On the method of studying a natural science, such as physiology (Edinburgh, 1894)

<u>Biographical Studies</u>

<u>The Lancet</u>, 25 February 1899, (London, 1899)
Biographisches Lexikon der Hervorragenden Aerzte, <u>4</u>, (London, 1929), p.933
Comrie, J.D., History of Scottish Medicine, <u>2</u>, (London, 1932), pp.608, 610, 693, 694, 713, 727
Boase, F., Modern English Biography, <u>6</u>, (London, 1965), p.519
DNB, (London, 1968), supplement

<u>Manuscript Locations</u>

EPH; ES; EU; LS

William Rutherford SANDERS (1828–1881)

Physician; Professor of Pathology at the
University of Edinburgh

Elected F.R.S.E.   3 January 1870

<u>Portraits, etc.</u>

Print: see Comrie,J.D., History of Scottish
    Medicine, <u>2</u>, (London, 1932) **
Portrait and Bust: (J. Hutchinson, RSA, 1881),
    University of Edinburgh

Born in Edinburgh on 17 February 1828, the son
of James Sanders, M.D., William received his
early education at the Edinburgh High School.
Because of his father's ill-health the family
moved to Montpelier, France, where William
finished his secondary education, after which
he studied at Montpelier University,
graduating Bachelier-es-Lettres in 1844. He
returned to Scotland the same year, and began studying medicine at Edinburgh University. He
graduated M.D. in 1849, receiving a gold medal for his thesis "On the anatomy of the spleen". He
studied further for two years at Heidelberg and Paris, before returning to Edinburgh to become
pathologist to the Royal Infirmary, as well as tutorial assistant in the University's clinical
classes. In 1853 he was appointed conservator of the Royal College of Surgeons' Museum, and did
much work to draw the attention of medical students, giving lectures to introduce the museum to
them. In 1855 he gave a six-month course of lectures in the extra-mural school of Edinburgh on
the Institutes of Medicine. Two years later he was appointed physician to the Royal Infirmary,
and in 1869, he succeeded William Henderson as Professor of Pathology in the University of
Edinburgh. As professor, his major contribution was the introduction of practical teaching into
the course, and he was regarded as an excellent clinical teacher and pathologist. He contributed
many articles to medical and scientific journals. He died at 30 Charlotte Square, Edinburgh on
18 February 1881.

<u>Honours</u> <u>and</u> <u>Distinctions</u>

1847–48   President of the Royal Medical Society of Edinburgh
1853   Fellow of the Royal College of Physicians of Edinburgh

<u>Publications</u> *

On the structure of the spleen   (Edinburgh, 1850)
Vertical hemiplegia of the palate (Edinburgh, 1865)
Case illustrating the supposed connection of alphasia with right hemiplegia and lesion of the
left frontal convolution of the brain (Edinburgh, 1866)
"The variation or vanishing of cardiac organic valvular murmurs" (1869)
Method of examining and recording medical cases   (Edinburgh, 1873)

<u>Biographical</u> <u>Studies</u>

DNB, (London, 1968)
<u>Proc.</u> <u>R.</u> <u>Soc.</u> <u>Edinb.</u>, <u>11</u>, (Edinburgh, 1882), pp.333–338
Comrie, J.D., History of Scottish Medicine, <u>2</u>, (London, 1932), pp.697–8
<u>The</u> <u>Times</u>, 19 February 1881, (London, 1881)
<u>Edinb.</u> <u>Med.</u> <u>J.</u>, <u>26</u>, (2), (Edinburgh, 1881), pp.939–49

<u>Manuscript</u> <u>Locations</u>

EU

James SANDERSON (1812-1891)

Surgeon

Elected F.R.S.E.   6 April 1863

<u>Portraits, etc.</u>

None known

Born at Dunbar on 21 May 1812, James received his early education at Dunbar Grammar School. He then studied medicine and graduated M.D. at Edinburgh University. His first appointment was as a surgeon on East India Company ships and he travelled to St. Helena, Bombay, China, Calcutta and Ceylon. In 1836 he became surgeon in the Madras Medical Service, and in the following year was with the artillery corps at St. Thomas's Mount. The following year, Lord Elphinstone appointed him the organiser of the medical department in connection with a system of convict labour. Having successfully carried this out, he became medical officer of the Neilgherries, where he stayed three years. In 1844 he was put on the Presidency medical staff as Port and Marine Surgeon, and was later made District Surgeon by the Governor of Madras. He subsequently became medical attendant to the Governor and his successors. In 1854 he was appointed Garrison Surgeon of Fort St. George, Madras, and went with Lord Harris as his medical attendant in his tours through the provinces, returning to Great Britain with him in 1859. The following year, he was appointed to the Governor's body- guard in Madras, and accompanied Sir William Dennison on his Presidency tours. He retired to Edinburgh in 1863 where he became active in various professional and scientific societies. While in India he had done much work for the horticultural gardens at Ootacamund. He died on 28 March 1891.

<u>Honours</u> and <u>Distinctions</u>

1872   Honorary Treasurer of Scottish Meteorological Society
1883   Honorary Treasurer to the Ben Nevis Observatory

<u>Publications</u>

None known

<u>Biographical</u> <u>Studies</u>

<u>Proc.</u> <u>R.</u> <u>Soc.</u> <u>Edinb.,</u> <u>20,</u> (Edinburgh, 1895), pp.49-50
<u>Edinb.</u> <u>Med.</u> <u>J.,</u> <u>36,</u> (2), (Edinburgh, 1891), p.1072

<u>Manuscript</u> <u>Locations</u>

None known

John SCOTT   (1797-1859)

Physician

Elected F.R.S.E.   21 January 1850

<u>Portraits, etc.</u>

None known

Born on 26 January 1797, the youngest son of
Rev. James Scott, of Benholme, Kincardine, John
entered Marischal College at the age of
thirteen. After four years' study he was
apprenticed with two Brechin surgeons, then
continued his medical training at the London
Hospital and Edinburgh University, graduating
M.D. in 1820. He was in the service of the East
India Company as ship's surgeon and surgeon's
mate for a few years. In 1824 he began medical
practice at Barnes, Middlesex, which became
very successful. In 1845, he was successful in
becoming Examining Physician to the East India
Company. He was considered a good administrator
and able physician, and in the Indian Mutiny he
behaved with calm efficiency in his task of
speedily despatching medical stores. He retired
to London's West End to a lucrative practice.
He died on 18 January 1859.

<u>Honours and Distinctions</u>

1845   Fellow of the Royal College of Physicians of Edinburgh
1846   Honorary M.A., Aberdeen University
1858   Fellow of the Royal College of Physicians of London

<u>Publications</u>

De Febre Biliosa Orientali   (Edinburgh, 1820)

<u>Biographical Studies</u>

Royal College of Physicians of London, Library Catalogue, (London, 1912), p.118
Crawford, D.C., History of the Indian Medical Service, 1, (1914), pp.516-7, 2, pp164-5
Munk, W., Roll of the Royal College of Physicians of London, 4, (London, 1955), p.97

<u>Manuscript Locations</u>

E

William SELLER   (1798-1869)

Physician and Botanist

Elected F.R.S.E.   4 March 1850

<u>Portraits, etc.</u>

Portrait: (Sir J. Watson Gordon, 1850), RCPE **

Born in Peterhead, Aberdeenshire, in 1798, the son of a merchant, William was first educated at Edinburgh High School, before going to Edinburgh University, where he graduated M.D. in 1821. His mother had been widowed when he was young, so to help her financially, he tutored younger students while at University, and later ran a boarding house for medical students. For many years he was lecturer in materia medica in the extra-mural school, an examiner in medical and general literature and Morison lecturer in mental diseases to the Edinburgh Royal College of Physicians. He was also editor of the 'Northern Journal of Medicine' and joint-editor of 'The Character of Medicine as an Art' (1857). For many years he was physician to the Royal Infirmary and the Royal Public Dispensary. Seller was a keen botanist, and was active in many scientific societies of Edinburgh. He died at Edinburgh on 11 April 1869 and was buried in the Dean Cemetery.

<u>Honours and Distinctions</u>

1836	Fellow of the Royal College of Physicians of Edinburgh
1841-48	Librarian of the Royal College of Physicians of Edinburgh
1848	President of the Royal College of Physicians of Edinburgh
1854	President of the Medico-Chirurgical Society of Edinburgh
1862	Awarded the Makdougall-Brisbane Prize, Royal Society of Edinburgh
1877	President of the Botanical Society of Edinburgh

<u>Publications</u> *

The principles of theory indispensable to sound observation in the practice of medicine (Edinburgh, 1841)
Examination of the views adopted by Liebig on the nutrition of plants (Edinburgh, 1845)
Observations on some plants obtained from the shores of Davis Straits, <u>Trans. Bot. Soc. Edinb.</u>, <u>2</u>, (Edinburgh, 1846), pp.215-223
Physiology at the farm in aid of rearing and feeding live-stock (Edinburgh and London, 1867)

<u>Biographical Studies</u> *

<u>Edinb. Med. J.</u>, <u>14</u>, (2), (Edinburgh, 1869), pp.1053-57
<u>Trans. Bot. Soc. Edinb.</u>, <u>10</u>, (Edinburgh, 1868-70), pp.202-3
<u>Proc. R. Soc. Edinb.</u>, <u>7</u>, (Edinburgh, 1872), pp.26-30
Comrie, J.D., History of Scottish Medicine, <u>2</u>, (London, 1932), p.630
Boase, F., Modern English Biography, <u>3</u>, (London, 1968), p.486

<u>Manuscript Locations</u>

E

Edward James SHEARMAN (1798–1878)

Physician

Elected F.R.S.E.   7 February 1870

<u>Portraits, etc.</u>

None known

Born in 1798 in Wington, Somersetshire, Shearman was educated at Edinburgh University and St. George's Hospital, graduating M.D. in 1841 from Jena. He then settled in Rotherham, Yorkshire where he practised for the next fifty years. For thirteen years he was surgeon to the Rotherham Dispensary. He was instrumental in founding the Rotherham Hospital, and became physician there in 1872, holding the post till death. He devoted his attention to the use of the microscope as a guide to the diagnosis and prevention of disease. At one stage he was made medical officer to the Rotherham Board of Health, but was dismissed when his microscopic investigations of the town water gave offence to the Board. He died at his home at Moorgate, Rotherham on 2 October 1878.

<u>Honours and Distinctions</u>

1841   M.D., Jena
1843   Member of the Royal College of Surgeons
1848   Member of the Medico-Chirurgical Society of London
1869   Member of the Royal College of Physicians of London
1869   Fellow of the Royal College of Surgeons of England

<u>Publications</u>

"The changes in the urine affected by disease, and the tests to distinguish them",
<u>The Lancet</u>, 1845, i, p.554
An Essay on the Properties of Animal and Vegetable Life (London, 1845)
Retrospective medical address on Disease of the Chest (n.p., 1848)
"Two cases of albumen in the urine in the same house, caused by drinking water contaminated with lead, one ending in fatal apoplexy, the other in universal anasarca, from whose skin large quantities of urea were collected whilst in the hot-air bath", <u>Practitioner</u>, <u>12</u>, 1874, (London, 1874), pp.226, 401
"Serious cases of haematuria to such quantity as to produce syncope, without albumen, constantly alternating with large deposition of uric acid." <u>Practitioner</u>, <u>14</u>, 1875, (London, 1875), p.275
Description of the Rotherham Hospital (Rotherham, 1877)

<u>Biographical Studies</u>

Plarr, V.G., Lives of the Fellows of the Royal College of Surgeons of England,
<u>2</u>, (Bristol and London, 1930), p.284
<u>Proc. R. Soc. Edinb.</u>, <u>10</u>, (Edinburgh, 1880), pp.14-15
Desmond, R., Dictionary of British and Irish Botanists and Horticulturists. (London, 1977), p.553

<u>Manuscript Locations</u>

None known

Sir John SIBBALD (1833–1905)

Commissioner in Lunacy for Scotland

Elected F.R.S.E.   15 January 1872

<u>Portraits, etc.</u>

Portrait: (Sir G. Reid), RCPE **
Photograph: see Eddington,A., Edinburgh and
    Lothians at the opening of the twentieth
    century (Edinburgh, 1904)

Born in Edinburgh on 24 June 1833, the eldest son of William Sibbald, Sir John received his early education at Merchiston School. He then studied medicine at Edinburgh University and the Paris Medical School, graduating M.D. from Edinburgh in 1854. In 1862 he was appointed Medical Superintendent of the Argyll District Asylum, remaining in this post till he was made Deputy Commissioner in Lunacy for Scotland, eight years later. In 1879 he was appointed Commissioner, a post which he retained till 1899. During this period he was also editor of the 'Journal of Mental Science', (1871-2), and Morison Lecturer on Mental Diseases to the Royal College of Surgeons of Edinburgh in 1877. The Edinburgh District Lunacy Board asked him to act as their medical adviser in the construction of the new asylum for the city. he never lived to see its completion, but had the satisfaction of seeing Aberdeen City Authorities erect a similar asylum for their insane poor. He died at 18 Great King St., Edinburgh on 20 April 1905.

<u>Honours and Distinctions</u>

1855–56   President of the Royal Medical Society of Edinburgh
1876   Fellow of the Royal College of Physicians of Edinburgh
1878–99   Commissioner in Lunacy for Scotland
1899   Knighted

<u>Publications</u> *

The cottage system and gheel   (London, 1861)
Clinical instruction in insanity, a necessary element in medical education   (Lewes, 1871)
The relief of the Poor in Edinburgh, 1904, (Glasgow, 1904)

<u>Biographical Studies</u> *

<u>Trans. Bot. Soc. Edinb., 23,</u> (Edinburgh, 1906), p.211
Eddington,A., Edinburgh and Lothians at the opening of twentieth century. (Edinburgh, 1904), p.237
Who Was Who 1897-1916, 1, (London, 1967), p.648
Britten, J. and Boulger, G.S., A biographical index of deceased British and Irish Botanists, second edition, (London, 1931), p.558
<u>Br. med. J.</u>, 1905, (London, 1905), p.1019
Royal College of Physicians of Edinburgh, Historical sketch of the Royal College of Physicians (Edinburgh, 1925), p.14
Royal College of Physicians of London, Library Catalogue (London, 1912), p.1138

<u>Manuscript Locations</u>

None known

Alexander Russell SIMPSON (1835-1916)

Professor of Midwifery at the University of
Edinburgh

Elected F.R.S.E. 6 February 1871

<u>Portraits, etc.</u>

Portrait: (T. Martine Ronaldson, 1910), RCPE
Photograph: see Comrie,J.D., History of Scottish
     Medicine, (London, 1932) **
Photograph: see Leyland,J., Contemporary medical
     men, (Leicester, 1888)

Born at Bathgate on 30 April 1835, the son of
Alexander Simpson, banker, Alexander was
educated at Bathgate Academy, and then studied
medicine at Edinburgh University, Montpelier,
Berlin and Vienna. He graduated M.D. from
Edinburgh University in 1856. In 1870 he became
Professor of Midwifery and Diseases of Women
and Children at Edinburgh University, and Dean of the Medical Faculty. He was Gynaecologist to
the University Ward in the Edinburgh Royal Infirmary, and Obstetric Physician at the Royal
Maternity and Simpson Memorial Hospital. He introduced a modified form of midwifery forceps, on
the principal by which axis-traction was introduced. He was a fellow of many obstetrical
societies, both at home and abroad, including the Glasgow Obstetrical and Gynaecological
Society, the American Gynaecological Society, and of the Obstetrical and Gynaecological
Societies of Berlin, Detroit, Italy, Leipzig, Moscow and Paris. He died in Edinburgh after a
motor-car accident on 6 April 1916.

<u>Honours and Distinctions</u>

1855-56   President of the Royal Medical Society of Edinburgh
1889   President of the Medico-Chirurgical Society, Edinburgh
1891-93   President of the Royal College of Physicians of Edinburgh
1906   Knighted
n.d.   President of the Royal Gynaecological Association

<u>Publications</u>

On the complete evacuation of the uterus after abortion (Edinburgh, 1876)
The treatment of fibroid tumours of the uterus (Edinburgh, 1878)
Basilysis; a suggestion for comminuting the foetal head in cases of obstructed labour
(Edinburgh, 1880)
Contributions to obstetrics and gynaecology (Edinburgh, 1880)
On axis-traction forceps (Edinburgh, 1880)
The relations of the abdominal and pelvic organs in the female (Edinburgh, 1881)
History of the Chair of Midwifery and the diseases of women and children in the University of
Edinburgh (Edinburgh, 1883)

<u>Biographical Studies</u>

Eddington, A., Edinburgh and the Lothians at the opening of the twentieth century.
(Edinburgh, 1904), p.237
Grant, Sir A., Story of the University of Edinburgh, <u>2</u>, (London, 1884), p.423
Comrie, J.D., History of Scottish Medicine, <u>2</u>, (London, 1932), pp.605, 685-6, 689, 712

<u>Manuscript Locations</u>

EPH; EU

John SMITH (1798–1879)

Physician

Elected F.R.S.E.   2 April 1866

<u>Portraits, etc.</u>

None known

Born in 1798, the son of a brassfounder and farmer of Edinburgh, John was educated at the local Heriot's Hospital. Its governor recommended him as an apprentice to Dr. George Wood, son of Dr Alexander Wood. He then continued his medical studies at Edinburgh University, graduating M.D. in 1822. He acted as Dr. Wood's assistant and partner, and at his death, succeeded to his practice. This included the management of the Saughton Hall Asylum for the Insane. He was also visiting physician to the Old Charity Workhouse and City Bedlam in Forrest Road. Smith was highly regarded for his work with the poor, and had a considerable reputation for his treatment of insanity. He lived in India Street and died on 4 February 1879.

<u>Honours and Distinctions</u>

1833   Fellow of the Royal College of Physicians of Edinburgh
1865   President of the Royal College of Physicians of Edinburgh

<u>Publications</u>

"An account of dysentry as it occurred in the Edinburgh Charity Workhouse during 1832 and 1833", <u>Edinb. Med. Surg. J.</u> (Edinburgh, n.d.)
"Cases of mental derangements terminating fatally, with the appearances disclosed by inspection", <u>Edinb. Med. Surg. J.</u> (Edinburgh, n.d.)

<u>Biographical Studies</u>

<u>Edinb. Med. J.</u> <u>24</u>, (2), (Edinburgh, 1879), pp.957–8
<u>Proc. R. Soc. Edinb.</u>, <u>10</u>, (Edinburgh, 1880), pp.353–4
Comrie, J.D., History of Scottish Medicine, <u>2</u>, (London, 1932), p.631
Historical sketch and laws of the Royal College of Physicians of Edinburgh, (Edinburgh, 1925), pp.9,25

<u>Manuscript Locations</u>

ES

James SPENCE   (1812-1882)

Professor of Surgery at Edinburgh University

Elected F.R.S.E.   5 February 1866

<u>Portraits, etc.</u>

Photograph: see Comrie, J.D., History of Scottish
     Medicine, (London, 1932) **
Photograph: see MSS. Cat., EUL
Portrait: (J. Irvine), location unknown

Born in South Bridge Street, Edinburgh, on 31
March 1812, the son of James Spence, a
merchant, James received his early education in
Galashiels and the High School of Edinburgh. At
the age of thirteen he entered Edinburgh
University with the aim of qualifying as an
army surgeon. During his studies he was
apprenticed to a firm of chemists in Princes
Street, but despite this interruption he
completed his medical education, gaining a diploma from the Edinburgh Royal College of Surgeons
in 1832. He was too young for a commission in the army, but went as a surgeon on two journeys to
Calcutta on board 'Protector', an East India Company ship, before returning to Edinburgh, where
he survived a typhus attack. He was next a University demonstrator of anatomy under Alexander
Monro (tertius) for seven years. On resigning this post, he became demonstrator in the extra-
mural school of anatomy at Surgeons' Square. He was a very skilful dissector and was especially
interested in the treatment of tracheotomy, herniotomy, urinary diseases and amputations, while
being essentially conservative in his outlook on surgical practices. He later lectured on
surgery at the High School Yards and Surgeons' Hall, until he was appointed Professor of Surgery
at the University of Edinburgh in 1864, holding this post till death. By his contemporaries he
was regarded as a great operating surgeon, and from 1870 onwards was recognized as the leading
surgeon in Scotland. He died at 21, Ainslie Place, Edinburgh on 6 June 1882, and was buried
there in the Grange Cemetery.

<u>Honours and Distinctions</u> *

1849   Fellow of the Royal College of Surgeons of Edinburgh
1864   Professor of Surgery in Edinburgh University
1865   Surgeon in Ordinary to the Queen in Scotland
1867-68   President of the Royal College of Surgeons of Edinburgh

<u>Publications</u> *

An inquiry into the anatomy of the par vagum and spinal accessory of the eighth pair of nerves.
(Edinburgh, 1842)
Remarks on some points connected with the ligature of arteries.   (Edinburgh, 1843)
Successful case of primary amputation at hip-joint; a case of spontaneous gangrene of the lower
extremity and a rare form of arterial lesion.   (Edinburgh, 1864)
Lectures on surgery, 3 editions, (Edinburgh, 1868)
Observations on the nature, symptoms and treatment of constricted or strangulated herniae,
reduced en bloc.   (Edinburgh, 1879)

<u>Biographical Studies</u>

Boase, F., Modern English Biography, <u>3</u>, (London, 1965), p.683
<u>Edinb. Med. J.</u> 28, (Edinburgh, 1882), pp.89-96
Comrie, J.D., History of Scottish Medicine, <u>2</u>, (London, 1932), pp. 587, 619, 622, 629, 670, 672

<u>Manuscript Locations</u>

EU; SaU

Robert SPITTAL   (1804-1852)

Physician

Elected F.R.S.E.   5 April 1841

<u>Portraits, etc.</u>

None known

Born in 1804, Dr. Robert Spittal graduated M.D.
in 1832 from Giessen. He then moved to
Edinburgh on being appointed Assistant
Physician to the Royal Infirmary there. Whilst
in this post, he became particularly interested
in research on the causes of the sounds of
respiration and of the heart, publishing his
first results in Edinburgh in 1830 under the
title "A treatise on auscultation, illustrated
by cases and dissections". Robert Spittal
lectured in medical acoustics in 1838 and in
medicine at the extra-mural school in 1839. He
was one of the earliest Edinburgh physicians
to introduce the method of Laennec into
practice. Spittal devoted the rest of his life
to perfecting and making use of Laennec's ideas
and theories, backing these up with some
extremely popular lectures. Robert Spittal died
on 7 April 1852.

<u>Honours and Distinctions</u>

1829   President of the Plinian Natural History Society
1833   President of the Royal Medical Society of Edinburgh
1834   Fellow of the Royal College of Physicians of Edinburgh
n.d.   Member of the Hunterian Medical Society of Edinburgh

<u>Publications</u>

"Observations on the natural history of the chamaeles vulgaris or common chameleon",
<u>Edinb. New Phil. J.</u>, <u>6</u>, (Edinburgh, 1829)
"Case of cyanosis; both ventricles opened into the aorta; pulmonary artery rudimentary and
impervious", <u>Edinb. Med. Surg. J.</u>, <u>44</u>, (Edinburgh, 1835)
"Experiments and observations on the sound of the heart", <u>Edinb. Med. Surg. J.</u>, <u>46</u>,
(Edinburgh, 1836)
"Experiments and observations on the cause of the sounds of respiration",
<u>Edinb. Med. Surg. J.</u>, <u>51</u>, (Edinburgh, 1839)
Case of aneurism of the arch of the aorta (1842)

<u>Biographical Studies</u>

Biographisches Lexikon der Hervorragenden Aerzte, <u>5</u>, (Berlin, 1929), pp.369-370
Comrie, J.D., History of Scottish Medicine, <u>2</u>, (London, 1932), p.628
Desmond, R., Dictionary of British and Irish Botanists and Horticulturalists, (London, 1977),
p.577

<u>Manuscript Locations</u>

EU

James STARK (1811-1890)

Physician; Superintendent of Medical Statistics
for Scotland

Elected F.R.S.E.   7 March 1842

<u>Portraits, etc.</u>

None known

Born in 1811, James Stark was the son of the
Edinburgh printer John Stark of Huntfield,
Biggar in Lanarkshire. James was educated at
the Universities of Edinburgh, Paris and Bonn.
He graduated M.D. at Edinburgh University in
1833 and later married the eldest daughter of
Mr Adam Black. After finishing his education,
he set up a large and popular practice at 21
Rutland Street, Edinburgh, and although not
entirely neglecting this, he devoted a good
deal of time to the collection and analysis of
vast amounts of data on mortality. It was largely due to Dr. Stark's proficiency with
statistics, that the government of the day recognised the need for a public department for the
collection of returns on births, marriages and deaths in Scotland. Hence when the post of
Statistician to the Registrar-General was initiated, Dr. Stark was appointed and he held this
position for many years. In this capacity, he carried out a simple analysis on data on
mortality available between 1780 and 1820. His conclusion was that the increasing death-rate
over this period was due to a combination of recurrent epidemics, physical and moral degradation
and Irish immigration. Eventually Dr. Stark retired from this post due to ill-health and died
at Underwood, Bridge of Allan, Stirlingshire on 2 July 1890.

<u>Honours and Distinctions</u>

1839   Fellow of the Royal College of Physicians of Edinburgh
n.d.   Fellow of the Royal Scottish Society of Arts

<u>Publications</u> *

Scarlet Fever, Edinburgh 1835-6   (Edinburgh, 1836)
On the signs of pregnancy during the earlier months of gestation (Edinburgh, 1842)
Researches on the brain, spinal chord and ganglia   (Edinburgh, 1844)
Chemical constitution of the bones of the vertebrated animals   (Edinburgh, 1845)
Inquiry into the probable cause of the continued prevalence and fatality of small-pox (Edinburgh, 1845)
Report on the mortality of Edinburgh and Leith (Edinburgh, 1846)
Inquiry into the sanitary state of Edinburgh and the rate of its mortality since 1780 (Edinburgh, 1847)
Vital statistics for Scotland (Edinburgh, 1851)

<u>Biographical Studies</u> *

Biographisches Lexikon der Hervorrangenden Aerzte, <u>5</u>, (Berlin 1934), p.395
Boase, F., Modern English Biography, <u>6</u>, (supplement) (London 1965), p.609
Royal College of Physicians of London, Library Catalogue, (London 1912), pp.1176-7

<u>Manuscript Locations</u>

E; EU; SaU;

John Lindsay STEWART (1831-1873)

Surgeon and Botanist

Elected F.R.S.E.   5 February 1872

<u>Portraits, etc.</u>

None known

Born in Fettercairn, Kincardineshire, on 13
December 1831, little is known of John's early
life. He studied medicine at Glasgow
University, and graduated M.D. in 1853. He then
joined the Indian Medical Service and in 1856
became assistant surgeon to the Presidency of
Bengal. The following year he was present at
the siege, assault and capture of Delhi,
receiving a medal for his services during that
event. He served for some time in the Punjab,
North India and undertook a journey with
Brigadier-General Chamberlain and an
expeditionary force to Waziristan in 1860. In 1861 he was appointed Civil Surgeon at Bijnour in
Rohilkand. Three years later he was employed to arrange a system of forest conservancy. During
the whole of his time in India, Stewart devoted every spare moment to his great interest,
botany. He studied the vegetation of the Terai and North-West Himalaya, the flora of the
Rohilkand forests and the outer valley between the Ganges and Sardah. He travelled to Sindh, to
Kashmir, the Upper Indus, Chenab and Sutlej rivers. In all these journeys he took copious notes,
and published the results of his researches in numerous papers. He contributed to the cause of
forest administration in India, establishing large fuel and timber plantations in the Punjab. In
1869 he returned to Britain, and spent his time at Kew preparing a Forest Flora of Northern and
Central India. He had to return to India before it was completed, however, and after this his
health began failing. He died at the Hill Sanitarium at Dalhousie, Mid-Lothian, on 5 July 1873,
and was buried under an oak tree in Dalhousie Cemetery.

<u>Honours and Distinctions</u>

1865   Fellow of the Linnean Society

<u>Publications</u> *

Notes on the flora of the country passed through by the expeditionary force under Brigadier-
General Chamberlain, against the Mahsood Wuzzeria, 1860   (London, 1862)
Memoranda on the Peshawar Valley, chiefly regarding its flora (Calcutta, 1863)
The forest flora of N.W. and Central India: a handbook of the indigenous trees and shrubs of those
countries (London, 1874)
List of the principle trees and shrubs of North India, with synonyms (Edinburgh, 1876)

<u>Biographical Studies</u> *

Cleghorn, H., Obituary Notice of Dr. John Lindsay Stewart, <u>Trans. Bot. Soc. Edinb.</u>, <u>12</u>,
(Edinburgh 1876), pp.31-33
Britten, J. & Boulger, J.S., A biographical index of deceased British and Irish Botanists (London, 1931),
p.288
Boase, F., Modern English Biography, <u>6</u>, (London, 1965), p.624

<u>Manuscript Locations</u>

E(IH)

Sir Thomas Grainger STEWART   (1837-1900)

Professor of the Practice of Physic at the
University of Edinburgh

Elected F.R.S.E.   2 January 1866

<u>Portraits, etc.</u>

Portrait: (Sir George Reid), RCPE
Photograph: see Comrie,J.D., History of Scottish
    Medicine, (London, 1932) **
Photograph: see Leyland,J., Contemporary medical
    men, (Leicester, 1888)

Born in Edinburgh on 23 September 1837, the son
of Alexander Stewart, decorator, Sir James was
first educated at the High School and then
studied medicine at the University of
Edinburgh, graduating M.D. in 1858. He
continued his medical studies at universities
and hospitals in Berlin, Prague and Vienna,
before returning to Edinburgh and becoming house-physician to John Hughes Bennett and Thomas
Laycock. In 1862 he was appointed pathologist to the Royal Infirmary, lecturer on pathology at
Surgeons' Hall, and physician to the Sick Children's Hospital. He was an incessant worker
writing many papers on such subjects as kidney conditions, dilation of the bronchi and acute
atrophy of the liver. In 1873 he lectured on the practice of physic in the extra-academical
school of medicine. Three years later he successfully applied for the Professorship of the
Practice of Physic at the University. He was an excellent lecturer and his lectures were
translated into French, German and Russian. He had one of the largest practices in consultation
work in Scotland. His leisure time was devoted to archaelogy and Scottish History, and to
translating literary works. He was one of the first in this country to draw attention to the
deep reflexes in neuritis (multiple neuritis). He died at Edinburgh on 3 February 1900, and was
buried in the Dean Cemetery

<u>Honours and Distinctions</u>

1882   Physician-in-Ordinary to Queen Victoria in Scoland
1887   Honorary M.D., Royal University of Ireland
1887   Honorary M.D., University of Dublin
1887   Honorary Fellow of the Royal College of Physicians of Ireland
1889-91   President of the Royal College of Physicians of Edinburgh
1892   Honorary Fellow of the College of Physicians of Philadelphia
1894   Knighted
1897   Honorary LL.D., Aberdeen University
1898   President of the British Medical Association at Edinburgh

<u>Publications</u> *

On the waxy oramyloid degeneration of the kidney (Edinburgh, 1861)
On acute yellow atrophy of the liver (Edinburgh, 1865)
On a case of acute atrophy of the kidneys and liver of a pregnant woman (Edinburgh, 1865)
On dilation of the bronchi, or bronchiectasis (Edinburgh, 1867)
A practical treatise on Bright's Disease of the kidneys (Edinburgh, New York, 1868, 1871)
Report on cases of cerebellar disease treated in the Royal Infirmary during the last three years
(Edinburgh and London, 1898)
Lectures on giddiness and on hysteria in the male (Edinburgh, 1898)

<u>Biographical Studies</u> *

DNB, (London, 1968), supplement.
<u>The Lancet</u>, 10 February 1900, (London, 1900), pp.412–415
<u>Edinb. Med. J.</u>, March 1900, (Edinburgh, 1900), pp.307–308
Comrie,J.D., History of Scottish Medicine, 2, (London, 1932), p.711
Logan-Turner,A.,(ed.), History of the University of Edinburgh, 1883-1933, (London, 1933), pp.128, 155
Biographisches Lexikon der Hervorragenden Aerzte, 5, (Berlin, 1934), p.425

<u>Manuscript Locations</u>

E; EU; LWM; O

John Addington SYMONDS   (1807-1871)

Physician to the General Hospital and Lecturer
on Forensic Medicine, Bristol

Elected F.R.S.E.   20 February 1854

<u>Portraits, etc.</u>

Bust: (Woolner), Private Collection
Photograph: (Unknown), NPG **

Born in Oxford on 10 April 1807, the son of
John Symonds, a medical practitioner, John
received his education at Magdalen College
School. He then went to study medicine at
Edinburgh University, graduating M.D. in 1828
with the thesis, "De Variola". He returned to
Oxford and assisted his father in his medical
work, before moving to Bristol in 1831, to
become physician to the General Hospital there.
He was also lecturer on forensic medicine at
the medical school in 1834, changing five years later to a lecturership on the practice of
medicine, which he held till 1845. Three years later he had to resign his hospital post because
of the pressures of his successful practice, and he was henceforth a consulting physician. In
1870 he was forced to retire because of failing health. His close friendship with Dr. James
Cowles Pritchard increased his interest in the problems of the insane. He was also interested in
the relations of mind and muscles, and the phenomena of dreams and sleep. He took a leading part
in the work of the British Medical Association in Bristol. He was also considered a good poet,
while much of his literary work on professional and other topics was published. He died on 25
February 1871.

<u>Honours and Distinctions</u>

1832-48   Leading physician to the General Hospital, Bristol
1836-45   Held leading post in the early affairs of the British Health Association
1857   Fellow of the Royal College of Physicians of London
1863   President of the British Medical Association, Bristol

<u>Publications</u> *

An account of the exhumation and appearance of a corpse, 14 months after death, and of the
detection of poisoning by arsenic (Worcester, 1835)
Age (London, 1835)
Sleep and dreams (London, 1851)
Habit: physiologically considered; a lecture .... at Bristol Literary and Philosophical
Institution (London, 1853)
The principles of beauty (London, 1857)
Remarks on criminal responsibility in relation to insanity (London, 1864)
Miscellanies (London, 1871)

<u>Biographical Studies</u> *

NUC, <u>580,</u> p.104
DNB, (London, 1968)
Symonds, J.A., Miscellanies, selected and edited, with an introductory memoir by his son
(London, 1871)
Tuke, Daniel Hack, Pritchard and Symonds in especial relation to Mental Science (London, 1891)

<u>Manuscript Locations</u>

E; MU; SaU

Sir Alexander TAYLOR (1790-1879)

Physician

Elected F.R.S.E. 5 January 1846

<u>Portraits, etc.</u>

None known

Born in Alton, Scotland, in 1790, the son of William Taylor, a shipowner, Alexander received his education at Edinburgh University, graduating M.D. in 1825 with the thesis, "De Febribus Malignis". He was staff surgeon to the English auxiliary force in Spain in 1835, where he was in charge of the medical hospital of the Casa de la Sociedad in Vittoria for a year. Three years later he settled in Pau, practising medicine there until his death. During the Franco-Prussian war of 1870-71, he was in charge of an ambulance for wounded soldiers. He made great efforts to develop the resources of Pau as a residence for invalids, and for this work, Napoleon III requested that Taylor be knighted. He was also made Knight of the Royal Crown of Prussia. He died at 5 Cayton Crescent, Hampstead, London, on 18 May 1879.

<u>Honours and Distinctions</u>

1865   Knighted (by Napoleon III)
n.d.   Knight of the Royal Crown of Prussia
n.d.   Member of the Historical Institute of France
1873   Vice-president of the Scientific Congress of France (at Pau)

<u>Publications</u>*

On the curative influence of the climate of Pau, and the mineral waters of the Pyrenees ..... (London, 1842)
Climates for invalids; with a description of the watering places of the Pyrenees and of the virtues of their respective mineral sources (London, 1856)
The general principles of medical climatology and their application to the climates of the south-east and south-west of France (London, 1875)

<u>Biographical Studies</u>

Boase, F., Modern English Biography, <u>3</u>, (London, 1963), p.897
<u>The Times</u>, 24 May 1879, (London, 1879)
<u>The Medical Times and Gazette</u>, <u>1</u>, 1879, (London, 1879), p.606
Biographisches Lexikon der Hervorragenden Aerzte, <u>5</u>, (Berlin, 1934), pp.524-5
List of graduates in medicine in the University of Edinburgh, from 1705 to 1866, (Edinburgh, 1867), p.77
NUC, <u>584</u>, p.227

<u>Manuscript Locations</u>

None known

Allen THOMSON (1809-1884)

Biologist; Emeritus Professor of Anatomy at the
University of Glasgow.

Elected F.R.S.E.   6 February 1843

<u>Portraits, etc.</u>

Portrait: (Sir D. Macnee, PRSA, 1877), Hunterian
    Museum, Glasgow.
Photograph: see Comrie, J.D., History of Scottish
    Medicine, (London, 1932) **

Born in Edinburgh on 2 April 1809, Thomson was
the second son of Professor John Thomson,
successively Professor of Military Surgery,
Medicine and General Pathology in the
University of Edinburgh. He received his early
education at the local high school, and then
went on to study medicine at Edinburgh
University and Paris. After graduating M.D.
from Edinburgh in 1830, he travelled to Holland and Germany to study anatomical and pathological
museums, before returning to Edinburgh to lecture on Anatomy and Physiology. He again visited
Europe in 1833 and 1837 making the aquaintance of some of the most famous scientific men of the
day. In 1837 he became private physician to the Duke of Bedford for two years, prior to taking
up the Chair of Anatomy at Marischal College, Aberdeen. In 1841, he resigned from this post due
to the small number of medical students at the University and became teacher of anatomy in the
extra-mural school at Edinburgh. The following year he was appointed Professor of the Institutes
of Medicine in the University of Edinburgh and six years later, to the Chair of Anatomy in the
University of Glasgow. During his time in Glasgow, Dr Allen Thomson gained much credit both as a
contributor to scientific literature and as a statesman, being instrumental both in the
completion of the Gilmorehill University Buildings and in the building of the Western Infirmary.
Dr Thomson retired from his professorship in 1877 and lived out the rest of his retirement in
London till his death on 21 March 1884. The cause of his death was a disease of the eyes which
showed symptoms of spreading to his brain just before he died.

<u>Honours and Distinctions</u> *

1871   LL.D., Edinburgh University
1871   President of the Biological Section of the British Association
1876   President of the British Association
1877   LL.D., Glasgow University
1878   Vice-president of the Royal Society of London

<u>Publications</u> *

Syllabus of lectures on physiology   (Edinburgh, 1835)
Outlines of physiology, for the use of students   (Edinburgh, 1848)
On the phenomena and mechanism of the focal adjustment of the eye to distinct vision at
different distances (Glasgow, 1857)
Structure and development of the brain; a lecture (London, 1879)

<u>Biographcal Studies</u> *

<u>Edinb. Med. J.</u>, <u>29</u>, (2), (Edinburgh, 1884), p.1157-62
Comrie, J.D., History of Scottish Medicine, <u>2</u>, (London, 1932), pp.: see index
<u>Proc. R. Soc.</u>, <u>13</u>, (London, 1887), p.xii
<u>Proc. Phil. Soc. Glasgow</u>, <u>15</u>, (Glasgow, 1884), pp.102-17

<u>Manuscript Locations</u>

E; EU; GC; L; LS;

William Burns THOMSON (1821-1893)

Medical Missionary

Elected F.R.S.E. 3 January 1870

<u>Portraits, etc.</u>

Print: see Thomson,W.B., Reminiscences of
    medical missionary work (n.p., 1895) **

Born in 1821 in Kirriemuir, Forfarshire,
Thomson's parents died when he was young. His
early education was in Golspie, Sutherland,
where his elder brother was the schoolmaster.
At the age of seventeen he decided to devote
himself to Christianity. He held a Sunday
School in the area, and the class grew from
three pupils to two hundred over the years. For
a time he was in charge of the school at
Golspie. In 1847 he went to Edinburgh to study
for the Ministry. During his college career
Thomson taught in Edinburgh schools and the Calton Jail to supplement his income. In 1854 he won
a prize from the Medical Missionary Society for his essay 'Medical Missions'. In 1859 he was
offered the position of residential superintendent at the Mission Dispensary, 39 Cowgate,
Edinburgh. At the time ill-health prevented him accepting the post, but eight months later he
was appointed. For the next eleven years Thomson did much to enlarge the Dispensary and increase
its influence, as well as establishing a training institute there. On 20 October 1865 Thomson
issued the first number of the 'Medical Missionary Journal'. In 1870 Thomson left the Missionary
Society's Dispensary, and set up his own premises in St. John Street, Canongate. In 1879
however, ill-health compelled him to leave Edinburgh for the South of France. He returned in
1881 and took up a post at Mildmay for eleven years. He retired from medical work and confined
himself to religious affairs. Thomson died on 23 April 1893, in Bournemouth, and was buried in
the cemetery there.

<u>Honours and Distinctions</u>

None known

<u>Publications</u>

Medical Missions (Edinburgh, 1854)
Look Before You Leap. A story for boys (Edinburgh, n.d.)
The City Arabs (London, 1868)
Reminiscences of Medical Missionary Work (London, 1895)

<u>Biographical Studies</u>

see <u>Publications</u>

<u>Manuscript Locations</u>

SaU

Sir John Batty TUKE (1835-1913)

Alienist and Medical Writer

Elected F.R.S.E.   2 February 1874

Portraits, etc.

Portrait: (G.P.Chalmers), RCPE **
Portrait: (D.Herdman, 1898), RCPE
Print: see Eddington,A., Edinburgh 'and Lothians
    at the opening of the twentieth century
    (Edinburgh, 1904)

Born in Beverley, Yorkshire, on 9 January 1835,
the eldest son of John Batty Tuke, Sir John was
educated at the Edinburgh Academy, and
Edinburgh University, where he graduated M.D.
in 1856. He then went to New Zealand as a
surgeon to the colonial forces, and became
senior medical officer in the Maori War for
three years from 1860. After his return to
Scotland his interest was mainly in the treatment of medical illness, and he soon attained
eminence in this field. In 1865 he was appointed medical superintendent of Fife and Kinross
Asylum, holding this post for eight years. After this he was in practice in Edinburgh,
specialising in mental illnesses and lecturing on. insanity. He was a Member of Parliament for
Edinburgh and St. Andrews Universities from 1900, as well as visiting Physician to the Saughton
Hall Private Asylum. He took an active interest in all educational matters, and was awarded
honorary degrees acknowledging both this and his professional eminence. He died on 13 October
1913.

Honours and Distinctions

1895   President of the Royal College of Physicians of Edinburgh
1896   Honorary D.Sc., Trinity College, Dublin
1898   Knighted
1912   Honorary LL.D., Edinburgh and St. Andrews Universities
n.d.   President of the Neurological Society of the United Kingdom

Publications *

Cases illustrative of the insanity of pregnancy, puerperal mania, and insanity of lactation
(Edinburgh, 1867)
On a new lesion observed in the brain (Edinburgh, 1868)
The cottage system of management of lunatics as practised in Scotland, with suggestions for its
elaborations and improvement (Edinburgh, 1869)
On the morbid appearances met with in the brain of thirty insane persons (Edinburgh, 1869)
Habitual drunkards (New York, 1893)
The insanity of over-exertion of the brain, ... the Morison lectures delivered before the
Royal College of Physicians of Edinburgh (Edinburgh and London, 1894)

Biographical Studies *

Eddington,A., Edinburgh and the Lothians at the opening of the twentieth century.
(Edinburgh, 1904), p.118
Who's Who 1913, (London, 1913)
Proc. R. Soc. Edinb., 34, (Edinburgh, 1914), p.9
Comrie, J.D., History of Scottish Medicine, 2, (London, 1932), pp.709, 713
Biographisches Lexikon der Hervorragenden Aerzte, 5, (Berlin, 1934), p.656

Manuscript Locations

SaU

Sir William TURNER (1832-1916)

Principal and Vice-Chancellor of the University
of Edinburgh

Elected F.R.S.E.   4 February 1861

<u>Portraits, etc.</u>

Portrait: (Sir J. Guthrie, 1913), University of
     Edinburgh
Portrait: (Sir G. Reid, 1895), location unknown
Photograph: see Comrie,J., History of Scottish
     Medicine, (London, 1932) **
Etching: (W. Hole, 1884), SNPG
Photograph: c.1875, MSS. Cat., EUL

William was born in Lancaster, England, on 7
January 1832, the son of a cabinetmaker, who
died when he was only five. He was brought up
by his mother, Margaret Aldren, receiving his
early education at a private school. At the age
of fifteen he was apprenticed to a local medical practitioner, and in the following year went to
St. Bartholomew's Hospital in London, qualifying as a member of the Royal College of Surgeons of
England in 1853, and graduating M.B. from London University four years later. On the
recommendation of Sir James Paget he became senior demonstrator to John Goodsir the Professor of
Anatomy at Edinburgh University, then the most distinguished School of Anatomy in Great Britain.
On Goodsir's death in 1867 he succeeded to the chair holding it for thirty- six years. From 1903
he was Principal and Vice- Chancellor of Edinburgh University. Earlier he had become interested
in the anatomy of cetacea, and published many papers on the subject. From 1870, he did much
research with regard to the placenta, and was also very interested in comparative and
anthropological anatomy, using his knowledge of craniology to support Huxley's defense of
evolutionary theory. He was critical of Dubois's work on Pithecanthropus erectus, which he did
not regard as a new species intermediate between man and ape but rather as an early evolutionary
form. In Edinburgh he was instrumental in the fund raising for the building of the new
university. His most lasting contribution was probably in medical education. He insisted on the
universities rights to grant their own medical degrees, while supporting the idea of joint
qualifying boards outside the universities, a concept which still forms the basis of the British
qualification in medicine. He began the 'Journal of Anatomy and Physiology' in 1867, and was its
editor for many years. Twenty years later he and George Murray Humphry founded the Anatomical
Society of Great Britain and Ireland. He was awarded many honours during his life, and received
honorary LL.D. degrees from Glasgow, St. Andrews, Aberdeen, Montreal, and Western University,
Pennsylvania, and was Honorary D.L.C. in Oxford, Durham and Toronto, to name a few. He died in
Edinburgh on 15 February 1916.

<u>Honours</u> <u>and</u> <u>Distinctions</u> *

1877   Fellow of the Royal Society of London
1882   President of the Royal College of Surgeons of Edinburgh
1886   Knighted
1892   President of the Anatomical Society of Great Britain
1898-1904   President of the General Medical Council
1900   President of the British Association for the Advancement of Science
1908   President of the Royal Society of Edinburgh

<u>Publications</u> *

On the employment of transparent injections in the examinations of minute structure of the human pancreas, further observations on the structure of nerve-fibres (London, 1860)
On the placentation of the sloths   (Edinburgh, 1873)
Atlas of human anatomy and physiology   (Edinburgh and London, 1876)
The convolutions of the brain; a study in comparative anatomy   (London, 1890)

<u>Biographical</u> <u>Studies</u> *

DSB, (New York, 1976)
Comrie, J.D., History of Scottish Medicine, <u>2</u>, (London, 1932), pp.621, 629, 653, 691, 692, 712
<u>Edinb.</u> <u>Med.</u> <u>J.</u> <u>16</u>, (Edinburgh, 1916), pp.169-70, 218-223; ibid, <u>30</u>, (Edinburgh, 1923), p.137
Who's Who, 1916, (London, 1916)

<u>Manuscript</u> <u>Locations</u>

EPH; EU; SaU; O

James WATSON (1792-1877)

Physician

Elected F.R.S.E. 4 January 1853

<u>Portraits, etc.</u>

None known

Born in Glasgow on 11 September 1792, James
received his early education at the Edinburgh
High School. He then studied medicine at
Edinburgh University, graduating M.D. in 1812.
He was appointed assistant surgeon to the East
India Company and lived for almost twenty years
in India. When he returned to Britain in 1831,
he retired from the company and settled in
Bath. Here he became the leading physician in
the city, and acquired a very large practice.
He devoted most of his time to hospital work,
and in later years concentrated on its
administrative and financial side. He died at
Bath on 27 September 1877.

<u>Honours and Distinctions</u>

1853 Fellow of the Royal College of Surgeons of Edinburgh

<u>Publications</u>

On idiopathic tetanus   (Glasgow, 1845)

<u>Biographical Studies</u>

<u>Proc. R. Soc. Edinb.</u>, <u>10</u>,   (Edinburgh, 1880), p.14
Crawford, D.G., History of the Indian Medical Service (London, 1914), p.361
Royal College of Physicians of London, Library Catalogue   (London, 1912), p.1301

<u>Manuscript Locations</u>

E;   EU

Morrison WATSON   (1845-1885)

Professor of Anatomy at Owen's College,
Manchester

Elected F.R.S.E.   3 March 1873

<u>Portraits, etc.</u>

None known

Born in Montrose in 1845, Watson received his
early education at the Edinburgh Institution.
He then studied medicine at Edinburgh
University, Berlin, and Vienna, concentrating
on anatomy and histology. In 1867, he graduated
M.D. from Edinburgh, his thesis being "On the
muscular anatomy of the posterior limb in
aves", which earned him a gold medal. For a
number of years he was assistant demonstrator
first with Professor John Goodsir, and then was
with Professor William Turner, until he was
appointed to the newly established Chair of Anatomy at Owen's College, Manchester, in 1874. He
also became Dean of the Medical Department in 1884 and held both posts until he died. He devoted
his professional career to improving the Manchester Medical School. Watson died on 25 March
1885.

<u>Honours and Distinctions</u>

1871   Member of the Royal College of Physicians of Edinburgh
1873   Fellow of the Royal College of Physicians of Edinburgh
1884   Fellow of the Royal Society of London

<u>Publications</u> *

Observations in human and comparative anatomy   (Edinburgh, 1874)
Notes of a case of double aortic arch   (London, 1877)
The homology of the sexual organs illustrated by comparative anatomy and pathology
(London, 1879)
Anatomy of the northern beluga (Edinburgh, 1879)
Report on the anatomy of the spheniscidae collected during the voyage of "H.M.S. Challenger",
(Edinburgh, 1886-95)

<u>Biographical Studies</u> *

Boase, F.,   Modern English Biography, <u>3</u>, (London, 1965), p.1228
<u>Edinb. Med. J.</u> <u>30</u>, (2), (Edinburgh, 1885), pp.1068-70
<u>Proc. R. Soc. Edinb.</u>, <u>14</u>, (Edinburgh, 1887), pp.131-3
Historical sketch and laws of the Royal College of Physicians of Edinburgh, (Edinburgh, 1925),
p.13
NUC, <u>651</u>, pp.93-4
List of the graduates in medicine in the University of Edinburgh, 1867-1900, (n.p., n.d.), p.146

<u>Manuscript Locations</u>

None known

Sir Patrick Heron WATSON (1832-1907)

Consulting Surgeon to Chalmers Hospital, Edinburgh

Elected F.R.S.E.   5 March 1866

<u>Portraits, etc.</u>

Photograph: see Comrie,J.D., History of Scottish Medicine, (London, 1932) **
Photograph: Medical Archives Centre, Royal Infirmary of Edinburgh

Born in Edinburgh on 5 January 1832, the son of the Rev. Charles Watson, minister of Burntisland, Sir Patrick received his early education at Circus Palace School and the Academy, Edinburgh. He then studied medicine in Edinburgh University, graduating M.D. in 1853. He was house-surgeon to Spence for a while, before joining the Army Medical Service and serving in the Crimea. Returning to Edinburgh, he became assistant to Professor Miller, succeeding eventually to the latter's surgical practice. In 1872 he began to lecture in clinical surgery when he became a surgeon to the Royal Infirmary. Later he became consulting surgeon to Chalmers Hospital and Leith Hospital. For twenty-five years he was a member of the General Medical Council and was appointed to the University Commission of 1889. During his life he held many honorary positions, including that of surgeon to Queen Victoria and King Edward in Scotland, and attache to the special embassy of the Earl of Roslin on the marriage of the King of Spain in 1878. He was Deputy Lieutenant of Edinburgh City, and had been awarded the Crimean, Turkish, French and Sardinian Medals, as well as the Jubilee, Royal Household and Coronation Medals, and by the order of Carlos III of Spain, the Insignia of Caballero. As a medical practitioner, he was considered pre-eminent in Scotland, both in the practice of surgery and medicine. With regard to the question of women entering medicine, he was of the opinion that they should be accepted on the same terms as men. He died in Edinburgh on 21 December 1907.

<u>Honours and Distinctions</u> *

1853	President of the Royal Medical Society of Edinburgh
1883-1901	Surgeon-in-Ordinary to the Queen in Scotland
1902-1907	Honorary Surgeon to the King in Scotland
1903	Knighted

<u>Publications</u> *

The modern pathology and treatment of venereal disease   (Edinburgh, 1861)
The radical cure of exomphalos in the adult   (Edinburgh, 1862)
Excision of the knee-joint: a description of a new apparatus for the after-treatment with illustrative cases (Edinburgh, 1867)
Case of strangulated hernia, perforation of the bowel, in which the occlusion of aperture by ligature was successfully adopted (Edinburgh, 1869)
Lithotrity and the new lever lithotrite (Edinburgh, 1869)

<u>Biographical Studies</u> *

Comrie, J.D., History of Scottish Medicine, 2, (London, 1932), p.712
Who's Who, 1907, (London, 1907)
Eddington,A., Edinburgh and Lothians at the opening of the twentieth century. (Edinburgh, 1904), p.241

<u>Manuscript Locations</u>

E; ES; EU; L

William WILLIAMS (1832-1900)

Principal and Professor of Veterinary Medicine
and Surgery, New Veterinary College, Edinburgh

Elected F.R.S.E.   3 February 1868

<u>Portraits, etc.</u>

Photograph: see <u>The Veterinary Journal</u>, <u>2</u>, 1900
Photograph: see Bradley, O.C., History of the
   Edinburgh Veterinary College, (London, 1923) **

Born in 1832 in Bantenwydd, Wales, William was
the fifth in direct line of a family who made
their livelihood mainly as farrier and cattle
doctors. At the age of seventeen he became a
pupil to a veterinary surgeon in Lancashire,
but three years later owing to ill health, he
left for Australia. In 1855 he returned with
his health restored, and completed his training
at the Dick Veterinary College, Edinburgh. He
had a brilliant career as a student, and he won the Highland Society's gold medal. After
finishing in 1857, Williams established an extensive practice in Bradford, Yorkshire. He was
appointed Principal of the Dick Veterinary College, Edinburgh in 1867, but before long friction
arose between the trustees, staff and students and he resigned in 1873. When he left the
majority of the students left with him, and set up the New Veterinary College in Gayfield
Square. He later moved to new premises in Elm Row (now the STV Gateway Studios). In 1904 his son
William Owen Williams moved the College to Liverpool, where it became part of the University.
Principal Williams held a prominent place among his colleagues. He was widely sought after as a
consultant at home and abroad and his reputation as a clinical adviser was unique. It was
Principal Williams who first demonstrated that 'ticks' conveyed disease to animals. He was for
some years editor of the 'Veterinary Journal' and his two great works, one on veterinary
medicine, the other on veterinary surgery were used in British and American Colleges. He was
involved with various veterinary medicine societies at home and abroad. Principal Williams ran
the New Veterinary College with great success until his unexpected death on 12 November 1900, at
his residence, 1, Crawford Place, Edinburgh. He had been married twice and left three sons and
two daughters. The number attending his burial ceremony, which took place in Warriston Cemetery,
demonstrated in a striking manner the high esteem in which the late Principal was held not only
by members of his own college, but by students, practitioners and public alike.

<u>Honours and Distinctions</u> *

1879   President of the Royal College of Veterinary Surgeons, Edinburgh

<u>Publications</u> *

The principles and practise of veterinary surgery   (Edinburgh, 1872)
The principles and practise of veterinary medicine   (Edinburgh, 1874)

<u>Biographical Studies</u> *

<u>Vet. J.</u> New series, <u>2</u>, 1900, (London, 1900), pp.299-315
<u>The Veterinarian</u>, <u>73</u>, 1900, (London, 1900), p.673
<u>Br. vet. J.</u>, 131, 1975, (London, 1975), p.141
Bradley,O.C., History of the Edinburgh Veterinary College, (London, 1923), pp.61-62, 65-68, 95-96
Boase,F., Modern English Biography, (London, 1965), p.1379

<u>Manuscript Locations</u>

EU

Thomas WILLIAMSON (1815-1885)

Surgeon at Leith Hospital

Elected F.R.S.E.   7 December 1857

<u>Portraits, etc.</u>

None known

Williamson was born at Greenock in 1815. At an early age he went to Edinburgh University to study medicine, graduating M.D. in 1836 with the thesis "On Empyema" and he also became a Licentiate of the Royal College of Surgeons of Edinburgh while yet only twenty. He held several public offices, and was for thirty years surgeon in Leith Hospital. For several years he was physician to Gladstone's Hospital for Incurables. In later years and up to the time of his death, he was parochial medical officer and vaccinator for the parish of South Leith, and medical officer of health for the whole of the burgh. He was an occasional contributor to medical journals on topics chiefly relating to sanitation. He published various pamphlets and delivered lectures on various aspects of public health. He was one of the oldest members of the Medico-Chirurgical Society, and of the Royal Society of Edinburgh. He died in Edinburgh of an attack of apoplexy on 30 December 1885.

<u>Honours and Distinctions</u>

1857   Fellow of the Royal College of Surgeons of Edinburgh

<u>Publications</u>

None known

<u>Biographical Studies</u>

<u>Proc. R. Soc. Edinb.</u>, <u>14</u>, (Edinburgh, 1887), p.167
<u>Edinb. Med. J.</u> <u>31</u>, Pt.II, (1886), p.796
List of the graduates in medicine in the University of Edinburgh, from 1705 to 1866, (Edinburgh, 1867), p.110

<u>Manuscript Locations</u>

None known

James George WILSON (1830-1881)

Professor of Midwifery at Anderson College,
Glasgow

Elected F.R.S.E.   16 February 1863

<u>Portraits, etc.</u>

None known

Born in 1830, the son of James Wilson, lecturer
on Midwifery in Portland Street Medical School,
Glasgow, James studied medicine at Glasgow
University, and graduated M.D. in 1853. On his
father's death in 1857, James succeeded to his
practice. He was physician-accoucheur in
Glasgow Maternity Hospital, before becoming
Professor of Midwifery at Anderson's College,
Glasgow in 1863, a post which he held until he
died. He lived for many years at 9 Woodside
Crescent, Glasgow and died there on 4 March
1881.

<u>Honours and Distinctions</u>

1854   Fellow of the Faculty of Physicians and Surgeons of Glasgow
1858   Fellow of the Royal College of Surgeons of Edinburgh
1865   Vice-President of the Obstetrical Society of London

<u>Publications</u>

Remarks on the postural treatment of prolapse of the funis; with cases illustrative of its
successful employment (Glasgow, 1867)

<u>Biographical Studies</u>

Boase, F., Modern English Biography, <u>3</u>, (London, 1965), p.1414
<u>Edinb. Med. J.</u>, <u>26</u>, (2), (Edinburgh, 1881), p.956
<u>The Lancet</u>, 1882, <u>1</u>, p.219
Comrie, J.D., History of Scottish Medicine, <u>2</u>, (London, 1932), p.661
<u>Trans. obstet. Soc. Lond.</u>, 24, 1883, p.44
NUC, <u>666</u>, p.689

<u>Manuscript Locations</u>

None known

Thomas Alexander WISE (1802-1889)

Principal of Dakka College, India

Elected F.R.S.E.   17 April 1854

<u>Portraits, etc.</u>

None known

Born on 13 June 1802, Thomas studied medicine at Edinburgh University, graduating M.D. in 1824 with the thesis "De Topographia Medica". Three years later he was appointed assistant surgeon with the Indian Medical Service in Bengal. He became Civil Surgeon at Hugli and while in this position, he founded Hugli College, and became its first Principal. He doubled the work of Principal with that of civil surgeon for three years from 1836, and was then made Secretary to the Committee of Education. His last post in India was as Principal of Dakka College. He retired on 11 February 1851. He died at Norwood on 23 July 1889.

<u>Honours and Distinctions</u>

1852   Fellow of the Royal College of Physicians of Edinburgh
1859   Fellow of the Royal College of Surgeons of England

<u>Publications</u>

Commentary on the Hindu system of medicine   (Calcutta, 1845)
Treatise on the diseases of the eye in Hindustan   (Calcutta, 1847)
The pathology of the blood   (Edinburgh, 1858)
Cholera   (Cork, 1864)
Review of the history of medicine,   2 vols.   (London, 1867)
History of paganism in Caledonia   (London, 1884)

<u>Biographical Studies</u>

Crawford, D.G., History of the Indian Medical Service, <u>2</u>, (London, 1914), pp.136-7, 153, 431, 438
Crawford, D.G.,   Roll of the Indian Medical Service   (London, 1930), p.60
Royal College of Physicians of London, Library Catalogue   (London, 1912), p.1332
Comrie, J.D.,   History of Scottish Medicine, <u>2</u>, (London, 1932), p.759

<u>Manuscript Locations</u>

None known

Alexander WOOD (1817-1884)

Physician

Elected F.R.S.E.   7 December 1863

<u>Portraits, etc.</u>

Portrait: (Sir J. Watson Gordon, 1861), RCPE
Caricature: see Kay, J., A series of portrait and
     caricature etchings, (Edinburgh, 1837)
Engraving: Medical Archives Centre, Royal
     Infirmary of Edinburgh **

Born in Cupar, Fife, on 10 December 1817, the
second son of Dr. James Wood, Alexander was
educated at a private school in Edinburgh run
by a Mr Hindmarsh, going to Edinburgh Academy
in 1826, where he stayed for six years. He then
entered Edinburgh University, and combined his
studies in the Arts with medical studies,
graduating M.D. in 1839. He became one of the
medical officers of the Stockbridge Dispensary in the year of his graduation, and afterwards of
the Royal Public Dispensary in the New Town. In 1841, he began lecturing on medicine in the
extra-mural school. He was closely connected with the Royal College of Physicians in Edinburgh,
and as well as holding various posts there, he represented the college on the General Medical
Council for fifteen years from 1858. As a General Medical Council representative he contributed
much to the improvement in sanitary reform for the poor. He was appointed assessor of the
Edinburgh University Court in 1864. Wood strongly opposed homeopathy and mesmerism. He was a
politician and philanthropist, a keen educationist and a Free-churchman. Wood is credited with
introducing the use of the hypodermic syringe into general practice. Wood married Rebecca Massey
on 15 June 1842. He retired from practice at the age of fifty-five, and died on 26 February
1884.

<u>Honours and Distinctions</u> *

1840   Fellow of the Royal College of Physicians of Edinburgh
1858   President of the Royal College of Physicians of Edinburgh

<u>Publications</u> *

Account of a case of poisoning with corrosive sublimate   (Edinburgh, 1830)
Homeopathy unmasked; being an exposure of its principal absurdities and contradictions, with an
estimate of its recorded cures (Edinburgh, 1844)
Sequel to homeopathy unmasked; being a further exposure of Hahnemann, and his doctrines
(Edinburgh, 1844)
What is mesmerism? An attempt to explain its phenomena on the admitted principles of
physiological and psychical science (Edinburgh, 1851)
"New method of treating neuralgia by the direct application of opiates to the painful points",
<u>Edinburgh Medical & Surgical Review, 82</u>, (Edinburgh, 1855), pp.265-281
Smallpox in Scotland   (Edinburgh, 1860)
Preliminary education (Edinburgh, 1868)

<u>Biographical Studies</u>

DNB, (London, 1968)
Comrie, J.D., History Scottish Medcine, 2, (London, 1932), pp.614-5
<u>Edinb. Med. J. 2</u>, (Edinburgh, 1884), pp.973-6

<u>Manuscript Locations</u>

L

Andrew WOOD (1810–1881)

Physician

Elected F.R.S.E.   21 December 1863

<u>Portraits, etc.</u>

None known

Born in 1810 in Edinburgh, the son of Dr. William Wood, and from a family of medical practitioners, Andrew received his early education at the Edinburgh High School. He went on to study medicine at Edinburgh University and while there he joined the Royal Medical Society. Later he was appointed Medical Officer of the New Town Dispensary. On his father's death Wood succeeded him in several important public appointments. He was surgeon to Heriot's Hospital, and also to the Merchant Maiden and Trades Maiden Hospitals. He held the position of Inspector of Anatomy for many years, and was for a time Manager of the Edinburgh Royal Infirmary. When the Edinburgh Medical Council was instituted, Wood was elected by the Royal College of Surgeons of Edinburgh as their representative, a post which he held until his death. During his later years, Wood produced many songs and writings on medical topics. He also wrote political and social articles for publication. Wood died on 25 January 1881, after suffering ill-health for some time.

<u>Honours and Distinctions</u>

1855-56   President of the Royal College of Surgeons of Edinburgh

<u>Publications</u>

None known

<u>Biographical Studies</u>

<u>Edinb. Med. J., 26</u>, Pt.II, (Edinburgh 1881), pp.854-61

<u>Manuscript Locations</u>

None known

Thomas WRIGHT (1809-1884)

Physician; palaeontologist; geologist

Elected F.R.S.E.   30 April 1855

<u>Portraits, etc.</u>

Photograph: see Leyland,J., Contemporary Medical
    Men, (Leicester, 1888) **

Born in Paisley, Renfrewshire, on 9 November
1809, Thomas received his early education at
the local grammar school. He was then
apprenticed to his brother-in-law, a local
surgeon, after that he went to study medicine
at the Royal College of Surgeons in Dublin, and
worked also at the Peter Street Anatomical and
Surgical School. He soon became a very skilled
dissector and knowledgeable anatomist, and he
was offered the post of demonstrator, which
could have led to a higher position.
Unfortunately he became so seriously afflicted with blood-poisoning after dissecting a case of
confluent smallpox, that he was unable to accept this position. On recovering, he qualified as a
surgeon in 1832, settling not long after at Cheltenham. Here he spent the rest of his
professional life, acquiring a large practice, and was for many years surgeon to the General
Hospital, and Medical Officer of Health to the urban district. In 1846, he graduated M.D. from
St. Andrews University. He was an enthusiastic researcher of palaeontology and geology. He
published thirty-two papers on geological subjects and made many contributions to the
Palaeontographical Society. He collected Jurassic fossils, and described sea urchins and star-
fish of Jurassic and Cretaceous formations. He died on 17 November 1884. Shortly after, his
collection of fossils was bought for an American museum. He had been married twice, and left 2
daughters and a son.

<u>Honours and Distinctions</u> *

1846   M.D., St. Andrews
1859   Fellow of the Geological Society
1875   President of the Geological Section of the British Association
1878   Wollaston Medallist
1879   Fellow of the Royal Society of London

<u>Publications</u> *

Monograph on the British fossil Echinodermata from the Cretaceous formations, 2v.,
(London, printed for the Palaeontographical Society, 1864-1908)
Monograph on the British fossil Echinodermata from the oolitic formations, 2v. in 8 parts
(London, printed for the Palaeontographical Society, 1855-80)
Monograph on the Lias Ammonites of the British Islands
(London, printed for the Palaeontographical Society, 1878-1886)

<u>Biographical Studies</u>

DNB, (London, 1968)
World Who's Who in Science (Chicago, [1968]), p.1828
Boase, F., Modern English Biography, <u>3</u>, (London, 1965), p.1520
NUC, <u>675</u>, p.463

<u>Manuscript Locations</u>

EU; L; LR

John WYLIE (1790-1852)

Physician-General in the Madras Army

Elected F.R.S.E.   2 February 1852

<u>Portraits, etc.</u>

Photograph: see Leyland, J., Contemporary Medical
    Men, (Leicester, 1888)
Engraving: (R.J. Lane), RCP **

Born on 20 May 1790, the son of George Wylie of
Glasgow, John studied medicine at Edinburgh
University and graduated M.D. in 1811 with the
thesis 'De Hepatitide'. The following year he
joined the Madras Army as assistant surgeon. In
1825 he was promoted to surgeon, and thirteen
years later, to Superintending Surgeon. He held
this post until 1846, when again he was
promoted, to Inspector-General of Hospitals,
and finally to the rank of Surgeon-General in
1851. He saw active service in the Dekkan War (1817-18) and played a gallant part in the battle
of Corygaum (1818), being mentioned in dispatches. He received the C.B. in August 1850 when the
Order of the Bath was first conferred on medical officers. He retired from service in 1851,
having received the award of a General Order of the Governor in Council, and left Madras to
reside in Scotland. He died suddenly at his home at Arndean, Dollar, Clackmannanshire on 16 June
1852.

<u>Honours and Distinctions</u>

1844   Fellow of the Royal College of Surgeons of England
1850   Companion of the Order of the Bath

<u>Publications</u>

None known

<u>Biographical Studies</u>

Boase, F., Modern English Biography, <u>3</u>, (London 1965), p.1535
Plarr, V.G., Lives of the Fellows of the Royal College of Surgeons of England, <u>2</u>,
(Bristol and London, 1920), pp.557-9
List of the graduates in medicine in the University of Edinburgh, from 1705 to 1866,
(Edinburgh, 1867), p.44

<u>Manuscript Locations</u>

EU

John YOUNG (1835–1902)

Professor of Natural History at the University of
Glasgow

Elected F.R.S.E.   2 February 1863

<u>Portraits, etc.</u>

Bronze relief: (N.C.F. Shannan), Hunterian Art
     Gallery, University of Glasgow
Photograph: see University of Glasgow Old and
     New... (ed.), W. Stewart, (Glasgow, 1891) **

Born on 17 November 1835 in Edinburgh, John
received his early education at the Edinburgh
High School, and then went on to study at
Edinburgh University, where he graduated M.D.
in 1857. He then worked as a physician in the
Edinburgh Royal Infirmary, and in the Edinburgh
Royal Asylum. In 1860 he joined the Royal
Geological Survey for six years. In 1864, he
slipped and injured his knee and, as a result, was left lame for the rest of his life. He was
appointed Professor of Natural History and Honyman Gillespie Lecturer on Geology at Glasgow
University in 1866, a position which he held until his death in 1902. He was the Keeper of the
Hunterian Museum, Glasgow University for many years.

<u>Honours and Distinctions</u>

1876   President of Section C of the British Association
1893   President of the Educational Institute of Scotland

<u>Publications</u>

Physical Geography   (London, 1874)

<u>Biographical Studies</u>

Who's Who, 1898, (London, 1898)
Comrie, J.D., History of Scottish Medicine, <u>2</u>, (London 1932), p.662

<u>Manuscript Locations</u>

EU;   L;   LR

SCOTLAND'S CULTURAL HERITAGE

<u>Previous volumes in the series of catalogues of
Fellows of the Royal Society of Edinburgh, elected 1783-1882</u>

<u>Volume I- One hundred Medical and Scientific Fellows of the Royal Society of
Edinburgh, elected 1783-1882</u>   <u>Published 1981</u>

This introductory volume to the series includes all 39 Fellows of the Physical
Class of the Society who were present at its first meeting on 3rd November
1783. The remaining 61 Fellows were selected from those elected during the
first fifty years of the Society's existence.

<u>Note</u>: It is regretted that this volume is completely sold out

<u>Volume 2- One hundred Literary Fellows of the Royal Society of Edinburgh,
elected 1783-1812</u>   <u>Published 1981</u>

The second volume to the series includes all Resident Fellows of the Literary
Class of the Society, of whom 57 were elected at its first meeting on 17th
November 1783. Some of the more well-known names included in this catalogue
are those of Sir Walter Scott, Adam Smith the political economist, and Henry
Dundas, the 1st Viscount of Melville. The volume is illustrated with several
fine etchings by John Kay.

<u>Volume 3- One hundred Medical Fellows of the Royal Society of Edinburgh,
elected 1783-1844</u>   <u>Published 1982</u>

This volume together with Volume 4 covers the majority of Medical Fellows
elected to the Royal Society of Edinburgh during its first hundred years. Many
of those included in this volume are well-known figures in medical history,
for example, the Edinburgh University Professors Sir James Young Simpson and
Sir Robert Christison. Inevitably some of the less well-known but often
equally important figures of the period are included. The volume reproduces
some of the fine portraits of the Fellows as well as some more of the Kay
etchings.

Volumes 2 and 3 are currently in print and are available at £5.25 each from
the address on the inside cover of this publication.

Volume 5 in the series will cover one hundred scientific and engineering
Fellows of the Royal Society of Edinburgh, elected 1783-1882. This publication
is being produced with generous financial sponsorship from B.P. Scotland, and
has a planned publication date of Spring 1983. The cost will be as for the
rest of the series, £5.25 per copy.

"With a high standard of presentation and economical laying-out of facts,
these make useful reference books"........

Review in 'Books in Scotland'   Winter 1982/83
